E1/

# Maths
# The Basic Skills

## Handling Data

June Haighton • Bridget Phillips

Published in 2004 by:
Nelson Thornes Ltd
Delta Place
27 Bath Road
CHELTENHAM
GL53 7TH
United Kingdom

05 06 07 08/10 9 8 7 6 5 4 3 2

A catalogue record for this book is available from the British Library

ISBN 0 7487 7864 0

Illustrations and Page make-up by Mathematical Composition Setters Ltd

Printed and bound in Great Britain by Antony Rowe

# About the Handling data Worksheet Pack

This Worksheet Pack contains **Activities and Worksheets** for students who are working on topics from the Handling Data section of the Adult Numeracy Core Curriculum at Entry levels 1 and 2 (E1 and E2).

Entry level 3 (E3), Level 1 (L1) and Level 2 (L2) are covered in the textbook '**Maths the Basic Skills. Curriculum Edition**'.

## Activities

This section includes **a game** and other activities. The instructions for these are grouped together at the start of the section with the corresponding card sheets following behind. In most cases you will need to photocopy a sheet onto card then laminate and cut out the pieces you need for the game or activity.

## Worksheets

These are arranged in order of topics as follows:

- Extract Information – Lists, Tables, Diagrams, Pictograms, Bar Charts
- Sort and Classify – One Criterion, Two Criteria
- Collect Information
- Represent Information – Lists, Tables, Diagrams, Pictograms and Bar Charts

Where a topic covers more than one level the worksheets are arranged in order of curriculum reference, starting with E1 before progressing to E2.

The curriculum element covered by a worksheet is indicated alongside the title of the worksheet and also on the sheet itself.

## Answers

Answers to the Worksheets can be found online at www.nelsonthornes.com/mathsthebasicskills

## How to use this Worksheet Pack

Use the Contents list to find out which Activities or Worksheets will help you to teach a particular topic or curriculum reference. The tabs on the side of the pages will help you to locate the sheet you need.

# Contents



## Activities



## Worksheets



 

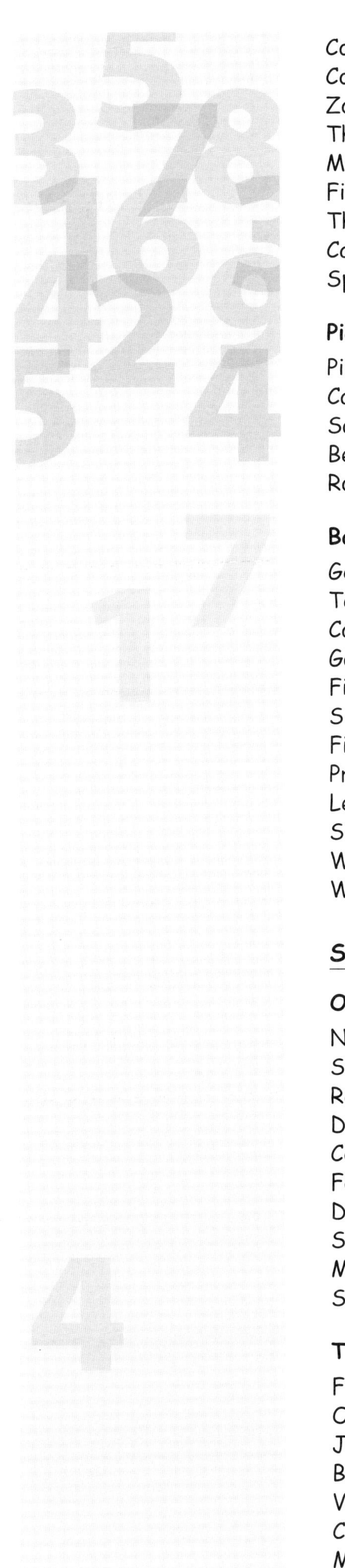

## Collect information


| | | |
|---|---|---|
| Record information | 71 | HD1/E2.4 |
| Travel | 72 | HD1/E2.4 |
| Takeaway | 73 | HD1/E2.4 |
| Favourite shop | 74 | HD1/E2.4 |
| Soaps | 75 | HD1/E2.4 |
| Questionnaire | 76 | HD1/E2.4 |
| Travel to work | 77 | HD1/E2.4 |


## Represent information

### Lists


| | | |
|---|---|---|
| Shopping lists | 78 | HD1/E1.3 |
| Decorating | 79 | HD1/E1.3 |
| Phone number list | 80 | HD1/E2.5 |
| Order the pages | 81 | HD1/E2.5 |


### Tables


| | | |
|---|---|---|
| Cleaning rota | 82 | HD1/E1.3 |
| Cricket | 83 | HD1/E2.5 |
| Timetable | 84 | HD1/E2.5 |


### Diagrams


| | | |
|---|---|---|
| Working in a shop | 85 | HD1/E1.3 |
| Market day | 86 | HD1/E1.3 |
| Design a garden | 87 | HD1/E1.3 |
| Library | 88 | HD1/E2.5 |
| Classroom plan | 89 | HD1/E2.1, HD1/E2.5 |
| Room plans | 90 | HD1/E2.5 |


### Pictograms and bar charts


| | | |
|---|---|---|
| Vending machine | 91 | HD1/E1.3 |
| Favourite cuppa | 92 | HD1/E1.3 |
| Keep fit | 93 | HD1/E1.3 |
| Weather | 94 | HD1/E1.3 |
| Watching TV | 95 | HD1/E2.5 |
| Jewellery | 96 | HD1/E2.5 |
| Surveys | 97 | HD1/E2.5 |
| Collect, record and display | 98,99 | HD1/E2.4, HD1/E2.5 |


 

# Instructions

N.B. Card sheets need to photocopied onto card, laminated and cut out before use.



## Pot of tea

HD1/E1.1

| NEEDED |
| --- |
| Pot of tea |

**WHAT TO DO** *(Individually or in pairs)*

Cut out the instruction cards and number cards separately. Discuss with students how to make a pot of tea and ensure they understand key words (eg brew).

Shuffle the instruction cards. Students put them in order and number them 1 to 8 using the number cards.

## List

HD1/E1.1

| NEEDED |
| --- |
| List cards |

**WHAT TO DO** *(Individually or in pairs)*

The cards in bold are group headings – place these face up in front of the students.

Shuffle the rest of the cards. Students take one card at a time and place it under the heading they think is most appropriate.

(Resulting groups may vary – promotes discussion of different ways of organising lists.)

## Recycling

HD1/E1.2

| NEEDED |
| --- |
| Recycling cards |

**WHAT TO DO** *(3 to 5 students)*

Based on the happy families game.

There are 5 sets – paper, compost, can, plastic and wood – each of 4 cards. Students' aim is to collect as many sets as possible. All the cards are shuffled and dealt. Each student looks at their hand of cards. The first player asks any other player if they have a particular card, e.g. 'Do you have compost 3?' If the player being asked has got the card they must hand it over. If they do not have it, the next player asks for a card. Whenever a player has a complete set they must say so, then that set is no longer asked for. The game ends when all the sets have been collected. The winner is the player with the most sets.

## Pot of Tea



Here are a list of instructions for making a pot of tea.
Put them in the right order and number them.

| | |
|---|---|
| Leave to brew | 1 |
| Pour water onto tea in pot | 2 |
| Throw water away | 3 |
| Boil the kettle | 4 |
| Pour a little boiling water in the tea pot | 5 |
| Swirl the water in the tea pot to warm the pot | 6 |
| Fill the kettle with cold water | 7 |
| Put tea in pot | 8 |

 

| Alphabetical | Random | By date or time | Numerical |
|---|---|---|---|
| Index | Shopping list | List of jobs | Bank statement |
| Phone book | Child's Christmas list | Time sheet | Diary |
| Yellow pages | Ingredients | Holiday schedule | Contents |
| Register | Invitation list | Address book | Set of instructions |
| Dictionary | Timetable | TV guide | List of appointments |
| Karaoke song list | Encyclopaedia | List of employees | Cinema guide |

  

## Recycling cards

HD1/E1.2

| | | | |
|---|---|---|---|
| 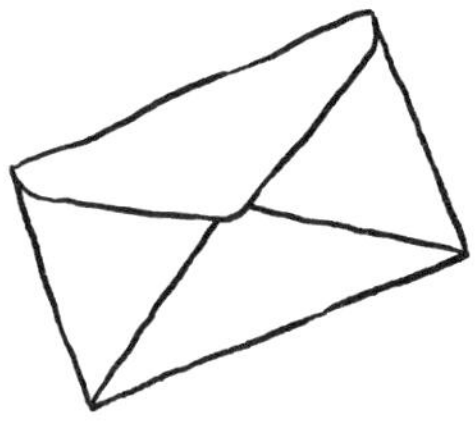<br>Paper 1 | 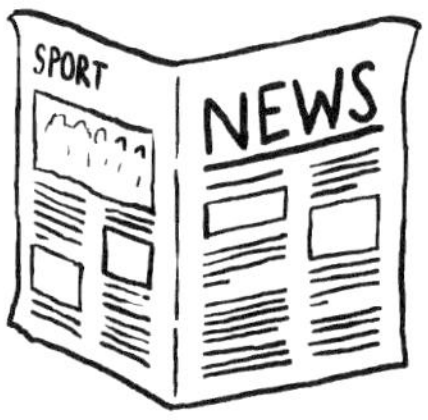<br><br>Paper 2 | 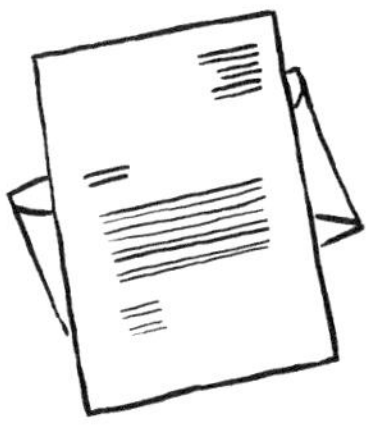<br>Paper 3 | 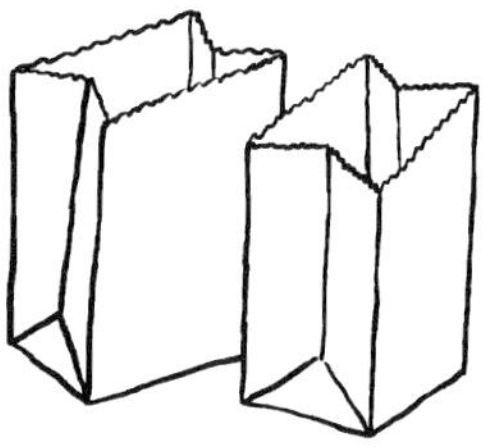<br>Paper 4 |
| 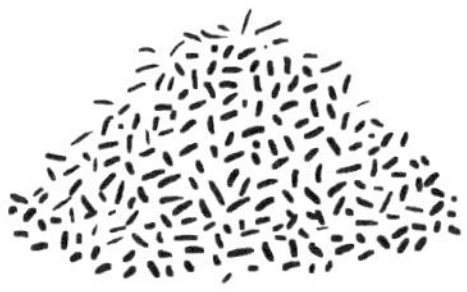<br>Compost 1 | 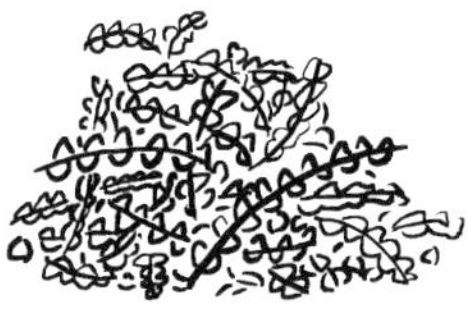<br>Compost 2 | 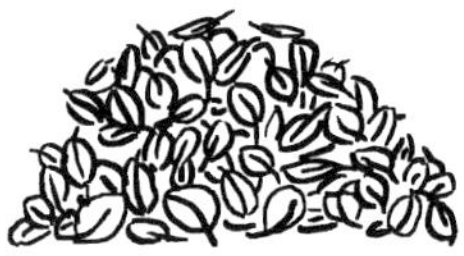<br>Compost 3 | 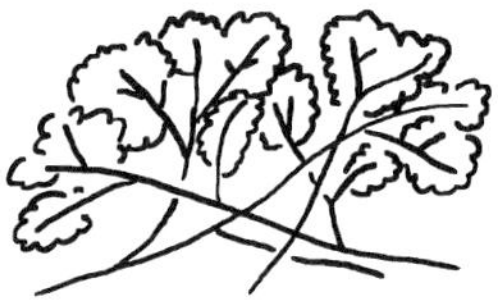<br>Compost 4 |
| 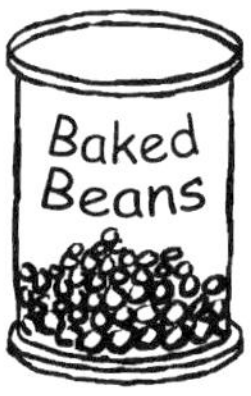<br><br>Can 1 | 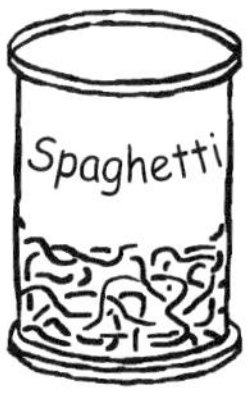<br><br>Can 2 | 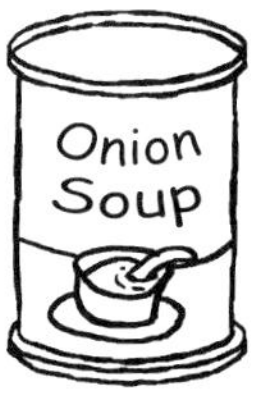<br><br>Can 3 | 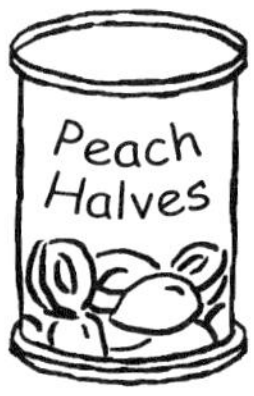<br><br>Can 4 |
| 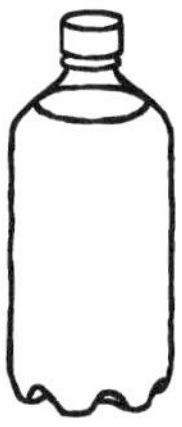<br>Plastic 1 | 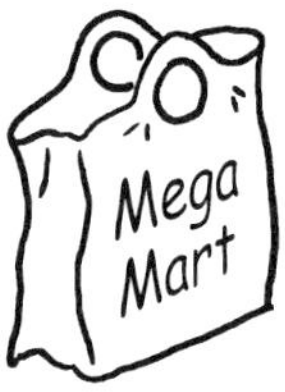<br><br>Plastic 2 | 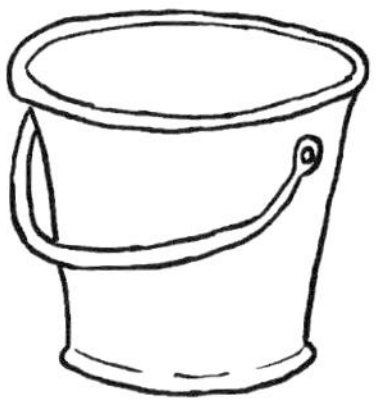<br>Plastic 3 | 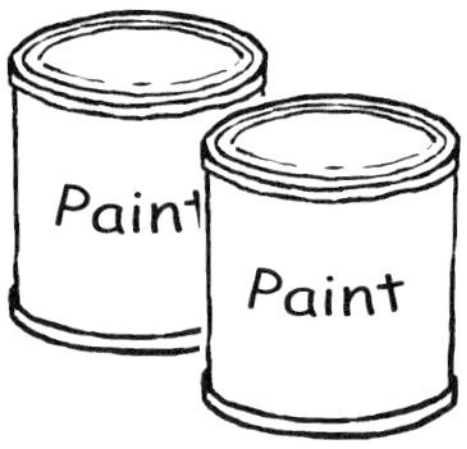<br><br>Plastic 4 |
| 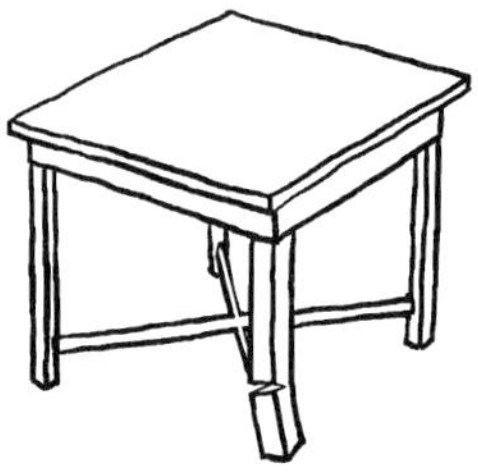<br>Wood 1 | 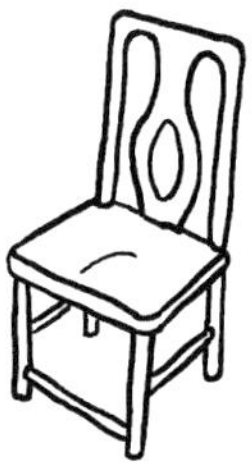<br>Wood 2 | 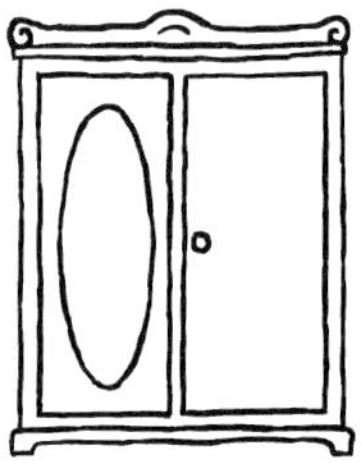<br>Wood 3 | 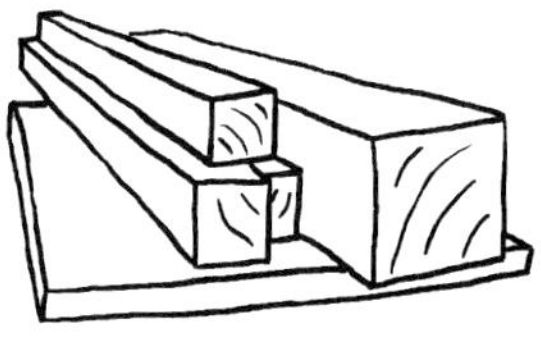<br>Wood 4 |

## Find a phone number



Lists are used to give information.

Lists can be arranged in different ways:

- **alphabetically** – in order of the alphabet like a phone book
- **numerically** – in number order
- **by date** – as in a bank statement
- **randomly** – not in any organised way, e.g. writing things on a shopping list as you think of them.

Here is a list of phone numbers.

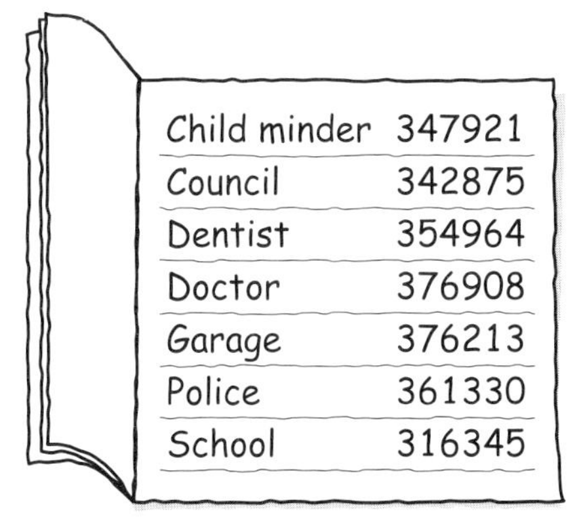

| | |
|---|---|
| Child minder | 347921 |
| Council | 342875 |
| Dentist | 354964 |
| Doctor | 376908 |
| Garage | 376213 |
| Police | 361330 |
| School | 316345 |



**1** Write the council's phone number below.

________________________________________

**2** Write the dentist's phone number below.

________________________________________

**3** Write the school's phone number below.

________________________________________

**4** Circle the word below that describes how this list is ordered.

randomly alphabetically by date numerically

## Corner shop

Below is a list of the corner shop opening times.

**Opening times**

| | |
|---|---|
| Monday | 9:00–5:00 |
| Tuesday | 10:00–5:30 |
| Wednesday | 9:00–12:00 |
| Thursday | 9:00–5:30 |
| Friday | 9:00–6:30 |
| Saturday | 10:00–4:00 |

1 When is the shop open on Thursday? ____________________

2 On which day is the shop only open in the morning? ____________________

3 On which day is the shop closed? ____________________

4 On which day does the shop close at 6:30? ____________________

5 What time does the shop close on Saturday? ____________________

6 On which day is the shop open from 9:00 to 5:00? ____________________

7 What time does the shop open on Tuesday? ____________________

8 What time does the shop close on Tuesday? ____________________

 

## Shopping list

Often in lists we show more than one item by writing a times sign (×) and then the number we want.

Here is a shopping list.
Use the list to answer the questions.

pint of milk x 4
loaf of bread x 3
jam
coffee x 2
tea
sugar
cake x 5

**1** How many different items are on the list?

______________________________

**2** What is the third item on the list?

______________________________

**3** How many jars of coffee are on the list?

______________________________

**4** How many loaves of bread are needed?

______________________________

**5** How many cakes are on the list?

______________________________

**6** How much milk is needed?

______________________________

**7** Is the list ordered alphabetically?

______________________________

## Holiday packing

HD1/E1.1

Here is a list of things to pack for a holiday.

Sunhat
Shampoo
T-shirt × 5
Conditioner
After sun
Flip flops
Toothpaste
Toothbrush
Bikini × 2
Sun cream
Shorts × 3
Magazine × 2
Sarong
Book × 3
Sunglasses

Worksheets

**1** Circle the word below that describes how this list is ordered.

alphabetically    numerically    by date    randomly

**2** What is the 5th item on the list? ______________________

**3** How many books are on the list? ______________________

**4** What is the 6th item on the list? ______________________

**5** What is the 10th item on the list? ______________________

**6** How many magazines are on the list? ______________________

**7** What is the 9th item on the list? ______________________

**8** How many bikinis are on the list? ______________________

 

## College notice board



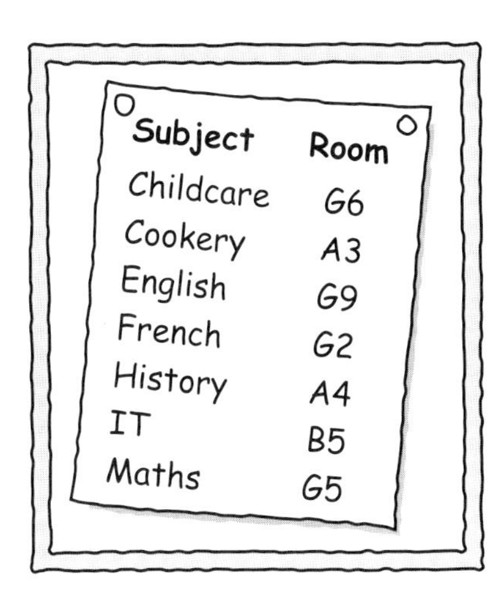

| Subject | Room |
| --- | --- |
| Childcare | G6 |
| Cookery | A3 |
| English | G9 |
| French | G2 |
| History | A4 |
| IT | B5 |
| Maths | G5 |

**1** Which room is Childcare in?

______________________________

**2** Where is IT?

______________________________

**3** Which subject is in G9?

______________________________

**4** Where is Maths?

______________________________

**5** Circle the word below which describes how this list is ordered.

alphabetically    numerically    by date    randomly

**6** Where is Woodwork?

______________________________

## Ordering lists



Lists are ordered alphabetically, numerically or by date so that you can find information quickly.

How do you expect the following to be ordered?

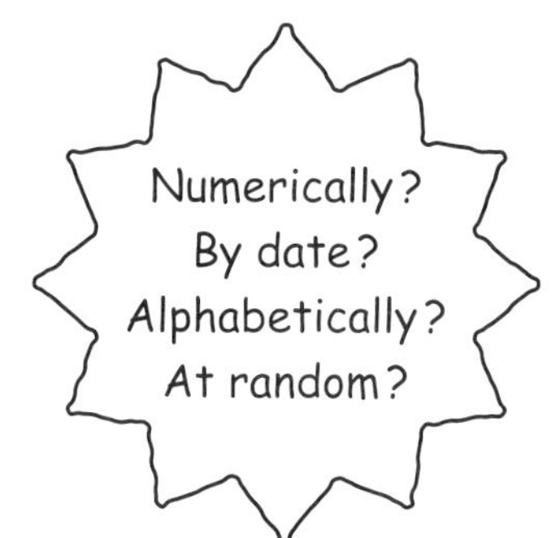


Appointments ______________________________

Phone numbers ______________________________

Term dates ______________________________

Birthdays ______________________________

A diary ______________________________

An index ______________________________

Child's Christmas list ______________________________

Yellow pages ______________________________

A list of instructions ______________________________



 

## Recipe



The recipe below is for Yorkshire puddings.

**Ingredients**

100 g plain flour
$\frac{1}{2}$ pint of milk
1 egg
2 tablespoons cold water

1 Preheat oven.
2 Heat oil in oven dish.
3 Place milk and flour in a bowl.
4 Mix together.
5 Add water.
6 Add egg and whisk.
7 Pour mixture into hot oven dish.
8 Cook for 20 minutes.



**1** How many ingredients are on the list? ____________________

**2** How many eggs do you need? ____________________

**3** How much milk do you need? ____________________

**4** Are the ingredients ordered alphabetically? ____________________

**5** What goes into the bowl with the flour? ____________________

**6** What is the last ingredient that you add? ____________________

**7** What do you pour the mixture into? ____________________

**8** How long do you cook it for? ____________________

  

## What is in it?

HD1/E2.1

Use the list on the side of this cereal box to answer the questions.

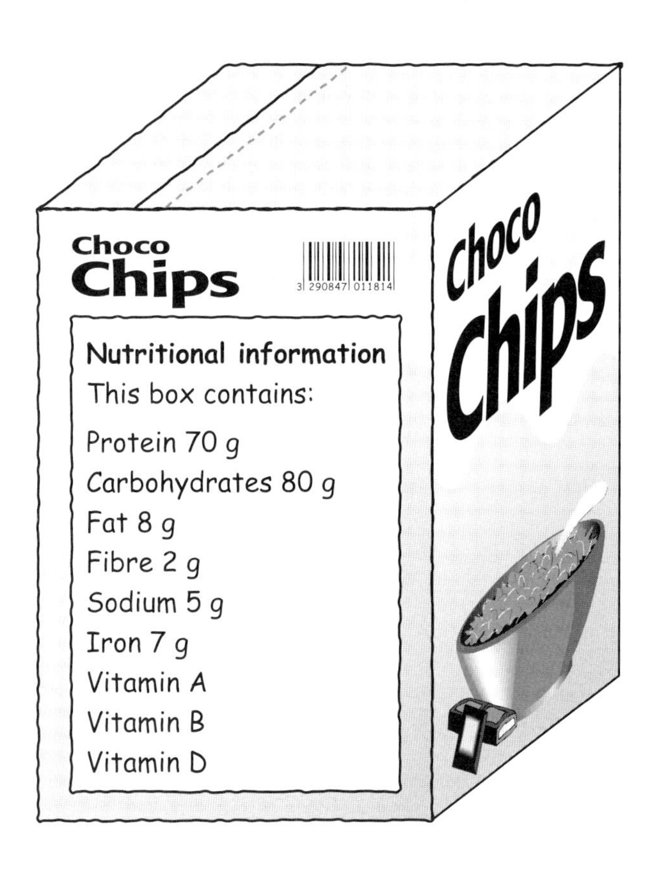


Does this cereal contain any fibre? ______________________

Which vitamins are in the cereal? ______________________

How much fat is in the cereal? ______________________

Does the cereal have any vitamin C? ______________________

There are 70 g of one item. Which is it? ______________________

Is there any sodium in the cereal? ______________________

Worksheets

 

## New authors



Jane has finished reading her book.
On the last page is a list of books by different authors.

| | |
|---|---|
| Alex Abbella | The Killing of the Saints |
| Susan Albert | Thyme of Death |
| George Batham | Sunset Moonrise |
| Jerome Charone | Maria's Girls |
| John Dale | Dark Angel |
| Stella Duffy | Calendar Girl |
| Stella Duffy | Eating Cake |
| Walter Mosely | Black Betty |
| Sam Reaves | Fear Will Do It |

Which author wrote more than one of these books? ______________________

How is the list ordered? ______________________

Which book did John Dale write? ______________________

Who wrote Black Betty? ______________________

Which book did Alex Abbella write? ______________________

Which author wrote Maria's Girls? ______________________

What is the title of Sam Reaves' book? ______________________

Which book did George Batham write? ______________________



  

## Christmas draw

The local pub has run a Christmas draw.
This table shows who bought the tickets.

| | | | | | | | | | |
|---|---|---|---|---|---|---|---|---|---|
| 1 Jim S | 2 Sam | 3 Mr Kirk | 4 Fred R | 5 Jim S | 6 Joe | 7 Jake | 8 Stan | 9 Tom P | 10 Bob B |
| 11 Julie P | 12 Shona | 13 Pat Grae | 14 Pat Grae | 15 Joe | 16 Ben | 17 J Wood | 18 Jane T | 19 Mrs Jones | 20 Mrs Watt |
| 21 Pat Grae | 22 Julie T | 23 Pat Grae | 24 Julie P | 25 Jim S | 26 Julie P | 27 Sam | 28 Nifa | 29 Vic | 30 Steph |
| 31 Lois | 32 Jim S | 33 Lois | 34 Colin | 35 Gill T | 36 Gill T | 37 Gill T | 38 Ian J | 39 Jim S | 40 Ian S |
| 41 Mr Kirk | 42 Ranjit | 43 Ranjit | 44 Pete | 45 Jim S | 46 Mog | 47 Olly | 48 Jack | 49 Jack | 50 Olly |

The following tickets have won prizes.

26 13 11 17 24 3

Who won with number 11? ______________________

Who won with number 17? ______________________

With which number did Mr Kirk win a prize? ______________________

Did Mrs Jones win a prize? ______________________

Who won the most prizes? ______________________

Who bought the most tickets? ______________________

Worksheets

 

## Football league



Below are some football league results.
Look at the table and answer the questions.

| Team | Matches played | Matches won | Matches lost | Matches drawn | Goals scored |
|---|---|---|---|---|---|
| Sandwich A | 4 | 4 | 0 | 0 | 8 |
| Broadstairs A | 4 | 2 | 1 | 1 | 6 |
| Westgate | 4 | 1 | 2 | 1 | 5 |
| Hartsdonians | 4 | 3 | 1 | 0 | 4 |
| Sandwich B | 4 | 2 | 2 | 0 | 5 |
| Broadstairs B | 4 | 3 | 0 | 1 | 6 |
| Broadstairs C | 4 | 4 | 0 | 0 | 7 |
| Ash | 4 | 1 | 2 | 1 | 5 |

How many games did Sandwich A win? ____________________

How many goals did Ash score? ____________________

Which teams won all their matches? ____________________

Did the Sandwich B team draw any matches? ____________________

Which team scored 7 goals? ____________________

How many teams drew a match? ____________________

Which team won 2 matches and lost 2 matches? ____________________

Which team scored the most points? ____________________

Which team scored the least points? ____________________



 

## College attendance



Below is a register.

| **Subject** Maths | | **Time** 10:00–12:15 | | | **Room** A209 | | | |
|---|---|---|---|---|---|---|---|---|
| **Last name** | **First name** | 17/10 | 24/10 | 31/10 | 7/11 | 14/11 | 21/11 | 28/11 |
| Abbott | Frank | / | A | / | / | / | / | / |
| Astley | Mary | / | / | / | / | / | A | / |
| Blake | John | / | / | / | / | A | / | / |
| Fripps | Joe | / | / | A | / | / | / | A |
| Hancock | Jane | A | / | / | / | / | / | / |
| Jackson | Keith | / | / | / | / | / | / | / |
| Maude | Connor | A | / | A | A | A | / | / |
| Shah | Hanifa | / | / | / | / | / | / | A |
| Tapikar | Mallika | / | / | A | / | / | / | / |

**1** What is the first name of the person called Fripps? ______________

**2** What is Keith's last name? ______________

**3** What is Mallika's last name? ______________

**4** Who was absent on the 24/10? ______________

**5** Who has attended all of the lessons? ______________

**6** Who has attended least? ______________

**7** On what date was John Blake absent? ______________

**8** On what date were there the most absences? ______________

**9** Circle the word which describes how the names are ordered.

alphabetically numerically randomly

# Banking

HD1/E2.1

Below is a page from a bank statement.
**Debit** means pay money out. **Credit** means pay money in.

| | Details | Debit | Credit | Balance |
|---|---|---|---|---|
| 23rd June | Brought forward | | | £172·00 |
| 24th June | 6565 Credit card Boots Canterbury | £6·41 | | £165·59 |
| 26th June | Cash withdrawal HSBC | £10·00 | | |
| 26th June | Cheque credit | | £12·00 | £167·59 |
| 27th June | Direct debit CIS Insurance | £7·80 | | £159·79 |
| 29th June | Standing order Greenpeace | £9·00 | | £150·79 |
| 2nd July | Charges | £15·00 | | £135·79 |
| 5th July | Cash withdrawal HSBC | £30·00 | | |
| 5th July | Cheque credit | | £50·00 | £155·79 |
| 6th July | Cheque 1007534 | £11·74 | | £144·05 |
| 9th July | 6565 Credit card Pizza Express | £45·91 | | £98·14 |
| 10th July | 6565 Credit card Virgin Record Store | £34·76 | | £63·38 |

1 How is this list arranged? ______________________________

2 How much was spent on 9th July? ______________________________

3 Was a credit or a debit received on 6th July? ______________________________

4 What was the balance on 23rd June? ______________________________

5 On what date was £7·80 spent? ______________________________

6 When was the debit for Pizza Express? ______________________________

7 On what date did the standing order go out? ______________________________



  

## Mileage chart



The chart below shows the distance from one town to another.
Look at the arrows to see how to use the chart.

**Example**: Manchester to Sheffield = **38 miles**

| | Birmingham | Brighton | Carlisle | Manchester | Middlesbrough | Newcastle | Norwich | Oxford | Sheffield |
|---|---|---|---|---|---|---|---|---|---|
| Brighton | 171 | | | | | | | | |
| Carlisle | 196 | 375 | | | | | | | |
| Manchester | 86 | 264 | 120 | | | | | | |
| Middlesbrough | 172 | 319 | 95 | 114 | | | | | |
| Newcastle | 202 | 359 | 59 | 145 | 39 | | | | |
| Norwich | 159 | 169 | 282 | 185 | 223 | 254 | | | |
| Oxford | 68 | 109 | 271 | 161 | 227 | 257 | 161 | | |
| Sheffield | 86 | 233 | 162 | 38 | 103 | 133 | 147 | 141 | |
| York | 129 | 276 | 117 | 72 | 50 | 88 | 181 | 184 | 60 |

The arrows go down from Manchester and across from Sheffield. The answer is where the arrows meet.

How far is it from Birmingham to Oxford? ______________

What is the distance from Sheffield to York? ______________

How many miles are there between Middlesbrough and Carlisle? ________

Which towns are 72 miles apart? ______________

What is the distance between York and Newcastle? ______________

Which towns are 59 miles apart? ______________

Which is nearer to Middlesbrough: Newcastle or York? ______________

Which two places are the same distance from Birmingham? ___________

 

# Town square

Which shop is between the supermarket and the cinema?

______________________________

What type of building is next to the butcher?

______________________________

Which shop is next door to the café?

______________________________

What is the name of the pub?

______________________________

What road is the café on?

______________________________

What is in the centre of the square?

______________________________



  

## Garden plan



Below is a diagram of a garden.
Some items in the garden are shown by symbols.
Use the key to find out what the symbols mean.
Then answer the questions.

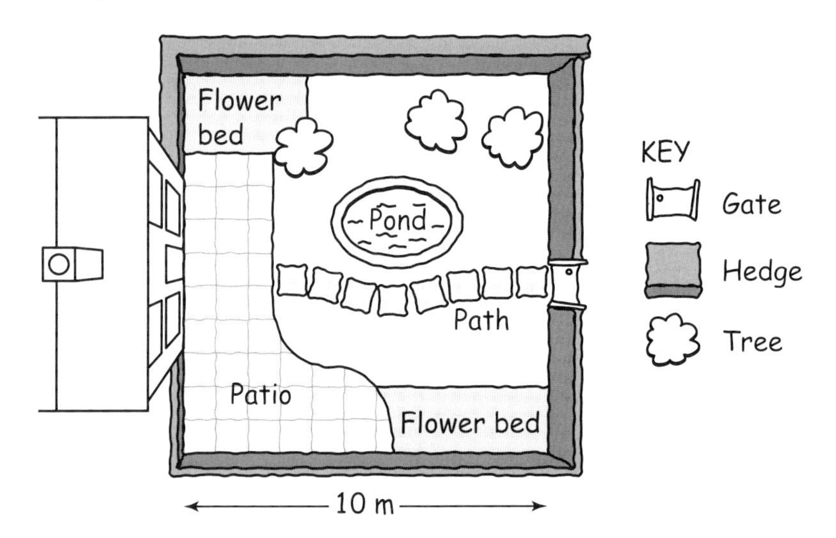


How many trees are there in the garden?

___

Is there a pond?

___

How long is the garden?

___

How many flower beds are there?

___

From the back gate where does the garden path lead to?

___

Does the garden have a fence or a hedge?

___





 

## Bedroom plan



Below is a plan of a bedroom.

All the measurements are in metres.

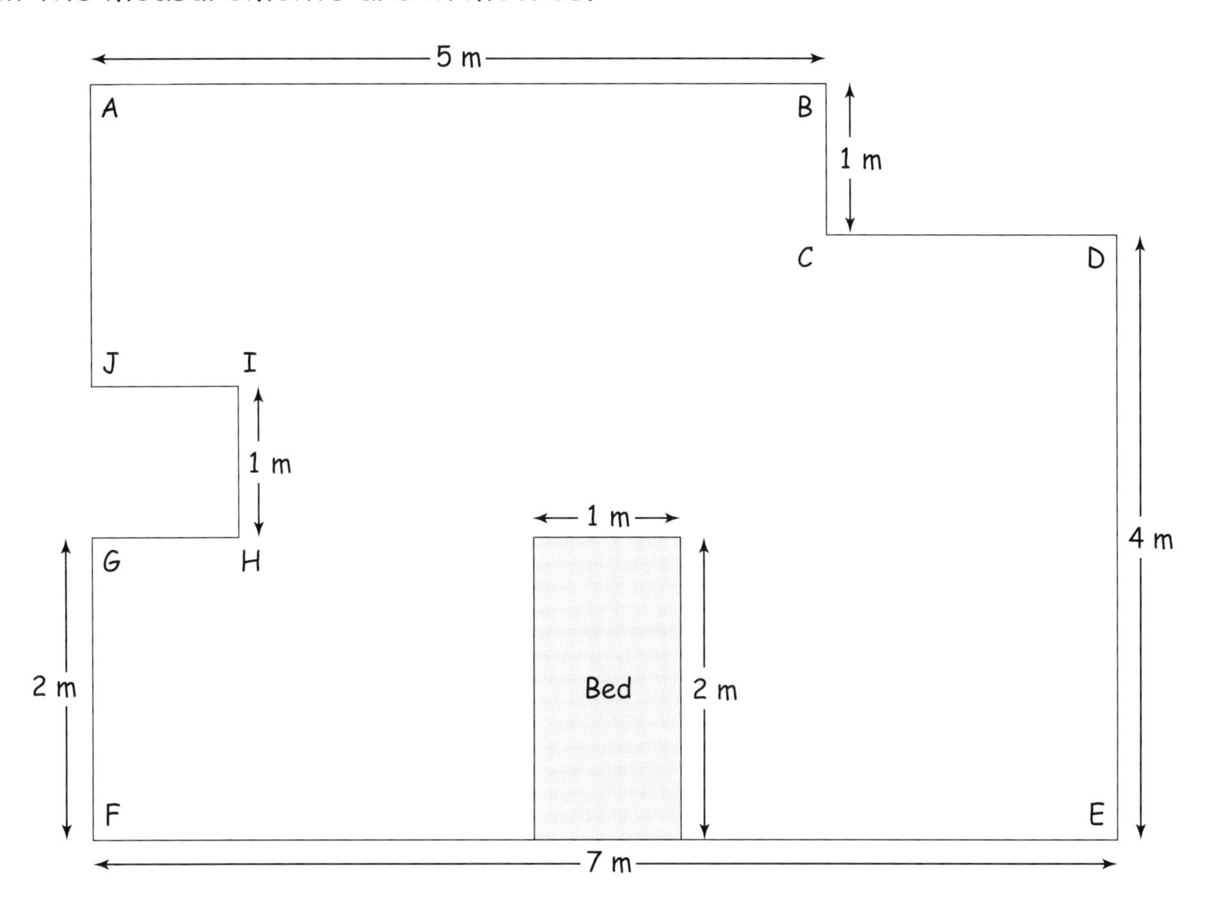


How long is the wall from F to E? ____________________

What is the distance between B and C? ____________________

What is the length HI? ____________________

How many metres is it from A to B? ____________________

Is it 6 m from G to F? ______________________________

How long is the bed? ____________________

How wide is the bed? ____________________

What is the distance from A to J? ____________________

What is the distance from C to D? ____________________

# Leisure centre plan



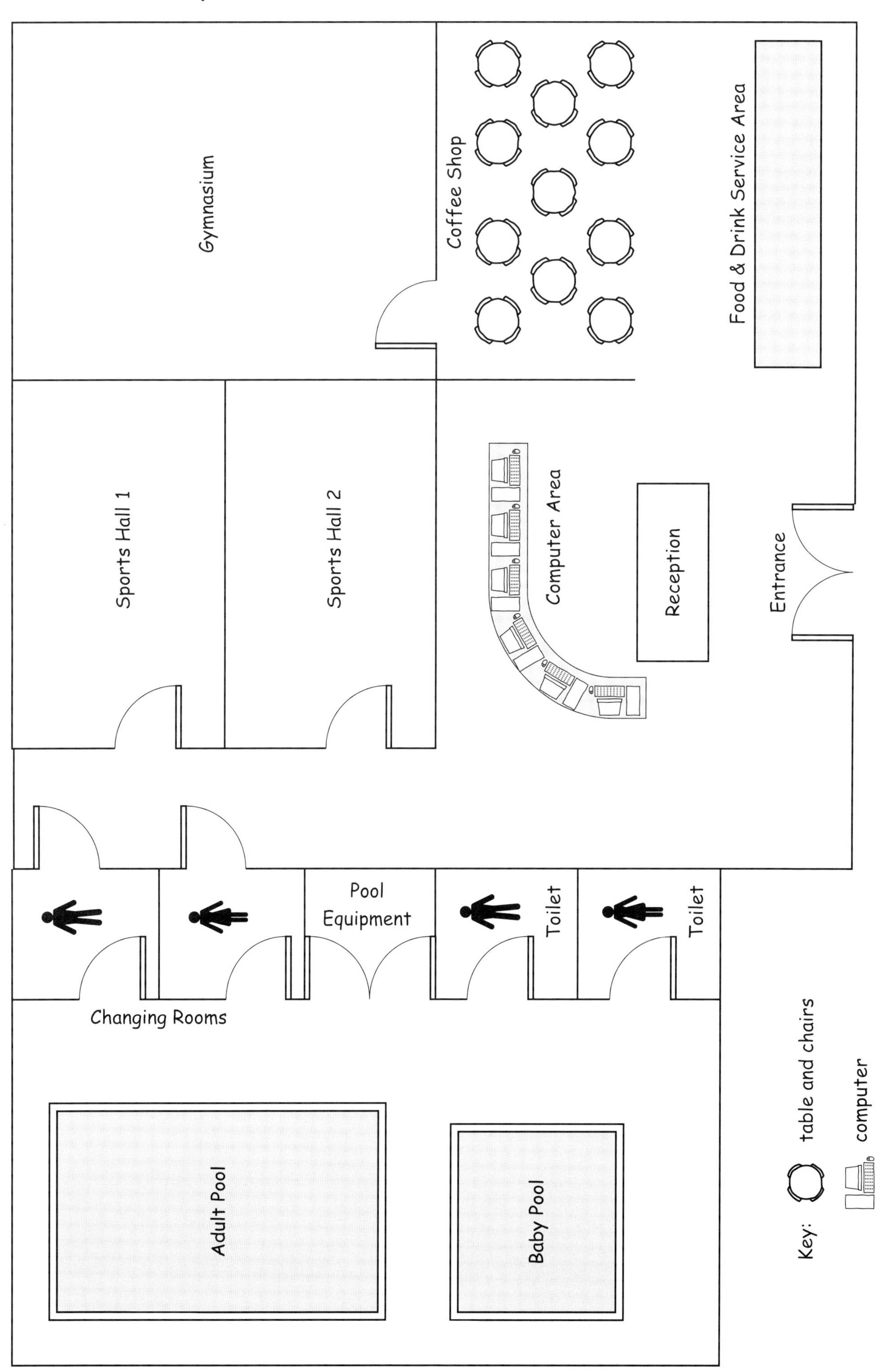




 

## Leisure centre questions

Answer these questions using the plan of the leisure centre.

How many sports halls are there?

______________________________________________

What is just inside the Entrance?

______________________________________________

What is between the toilets and the changing rooms?

______________________________________________

How many tables are there in the coffee shop?

______________________________________________

You go in the main entrance and turn right. Where are you?

______________________________________________

How many computers are there?

______________________________________________

How do you get to the gymnasium?

______________________________________________

How do you get to the pools?

______________________________________________

# Hospital plan





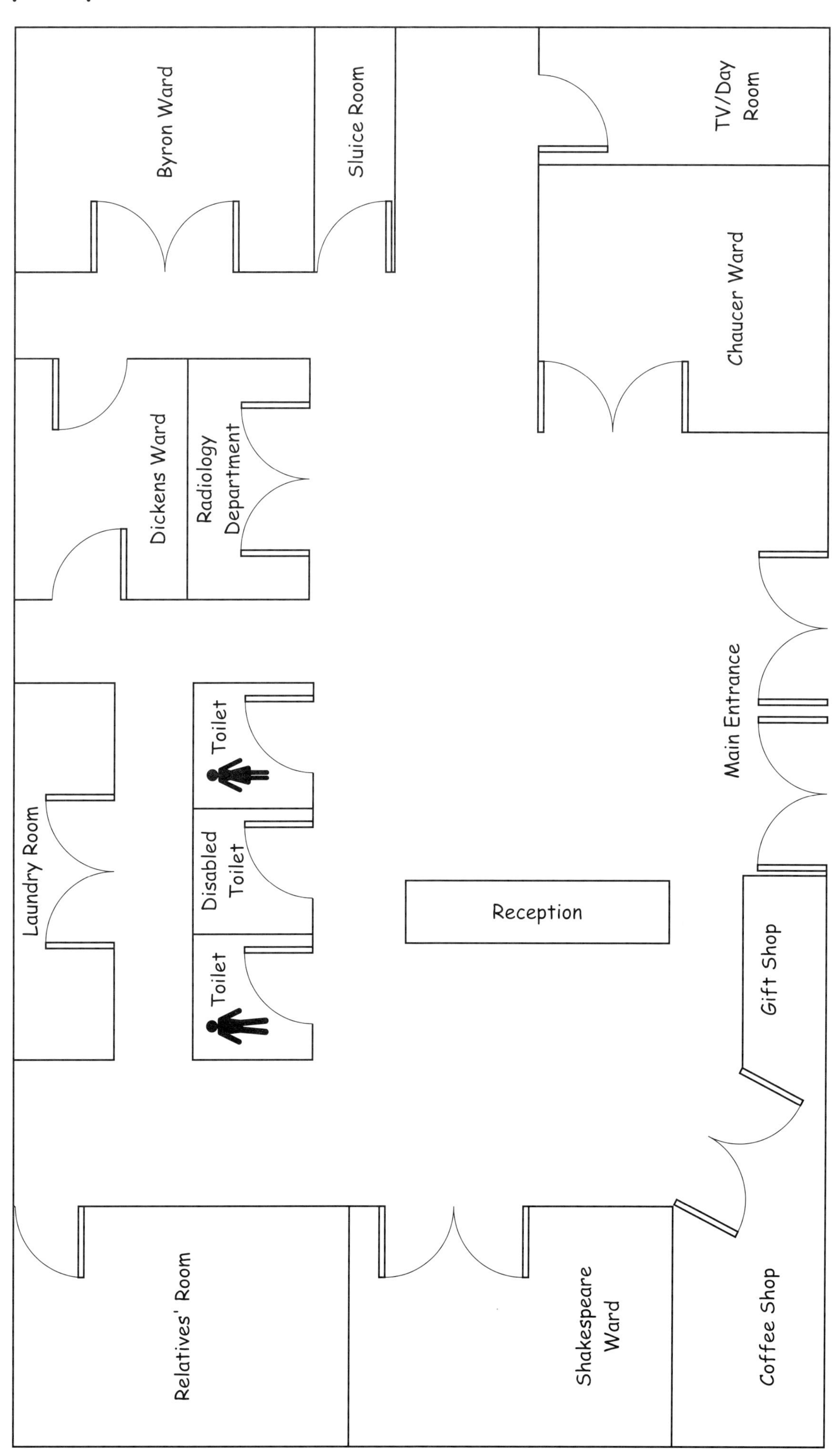




 

# Hospital questions



Use the plan of the hospital to answer these questions.

From the main entrance how do you get to Chaucer Ward?

______________________________________________

Which ward is between the Relatives' Room and the Coffee Shop?

______________________________________________

You come in through the left-hand doors of the main entrance.

What is straight in front of you?

______________________________________________

What is between the ladies' toilet and the men's toilet?

______________________________________________

Which room is behind the toilets?

______________________________________________

The Sluice Room is next to which ward?

______________________________________________

Which ward backs onto the Radiology Department?

______________________________________________



  

# College plan

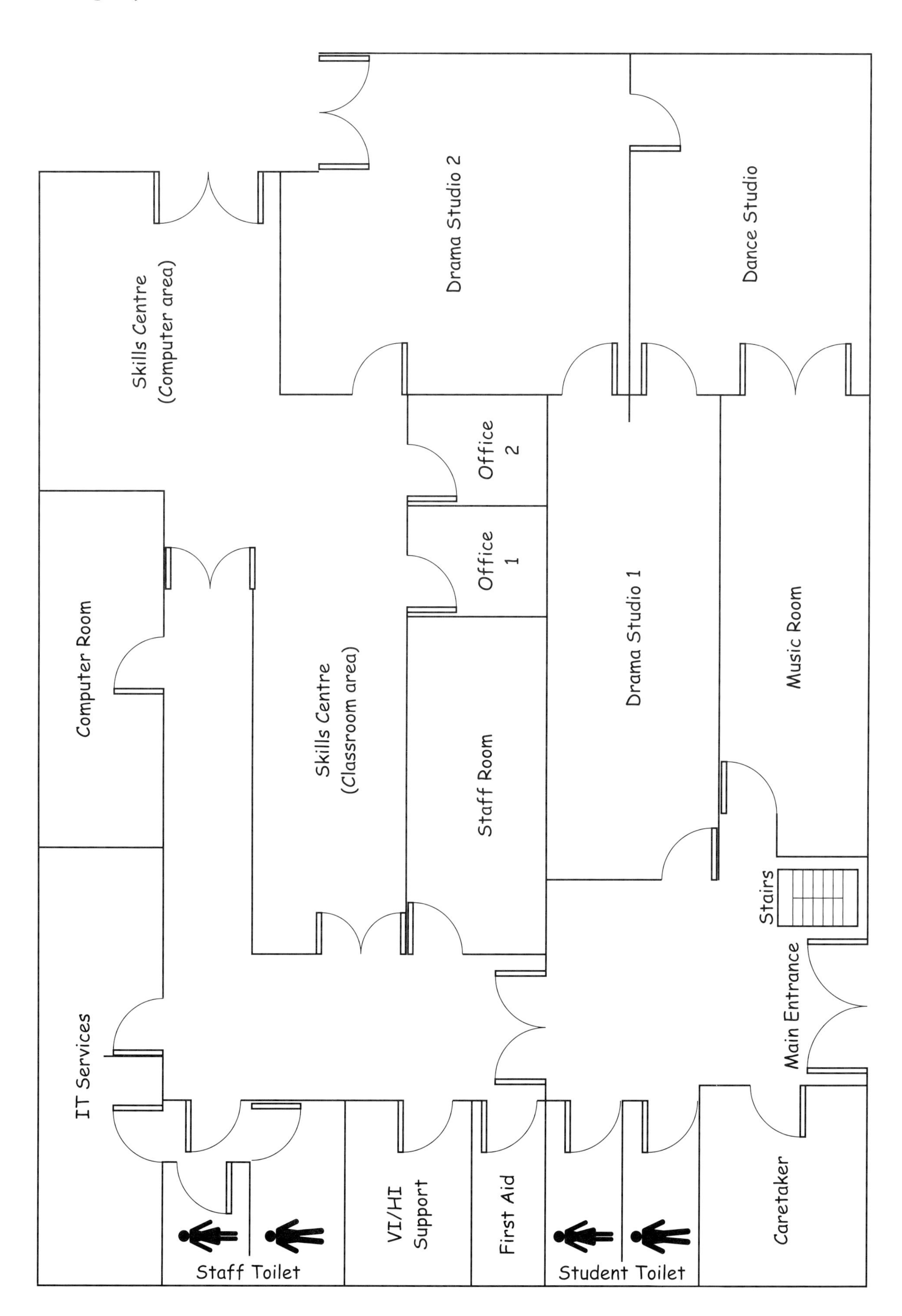






 

## College plan questions



Use the plan of the college to answer these questions.

What is to the left of the main entrance?

________________________________________

Which room are the stairs next to?

________________________________________

If you walk from the entrance through the Music Room and out of the other door, where will you be?

________________________________________

Come out of the Staff Room and go into the room across the corridor.

Where are you?

________________________________________

Which rooms are on either side of the VI/HI Support room?

________________________________________

Come out of Office 2, turn right and enter the room. Where are you?

________________________________________

Which rooms are on either side of the Computer Room?

________________________________________



  

## College campus map

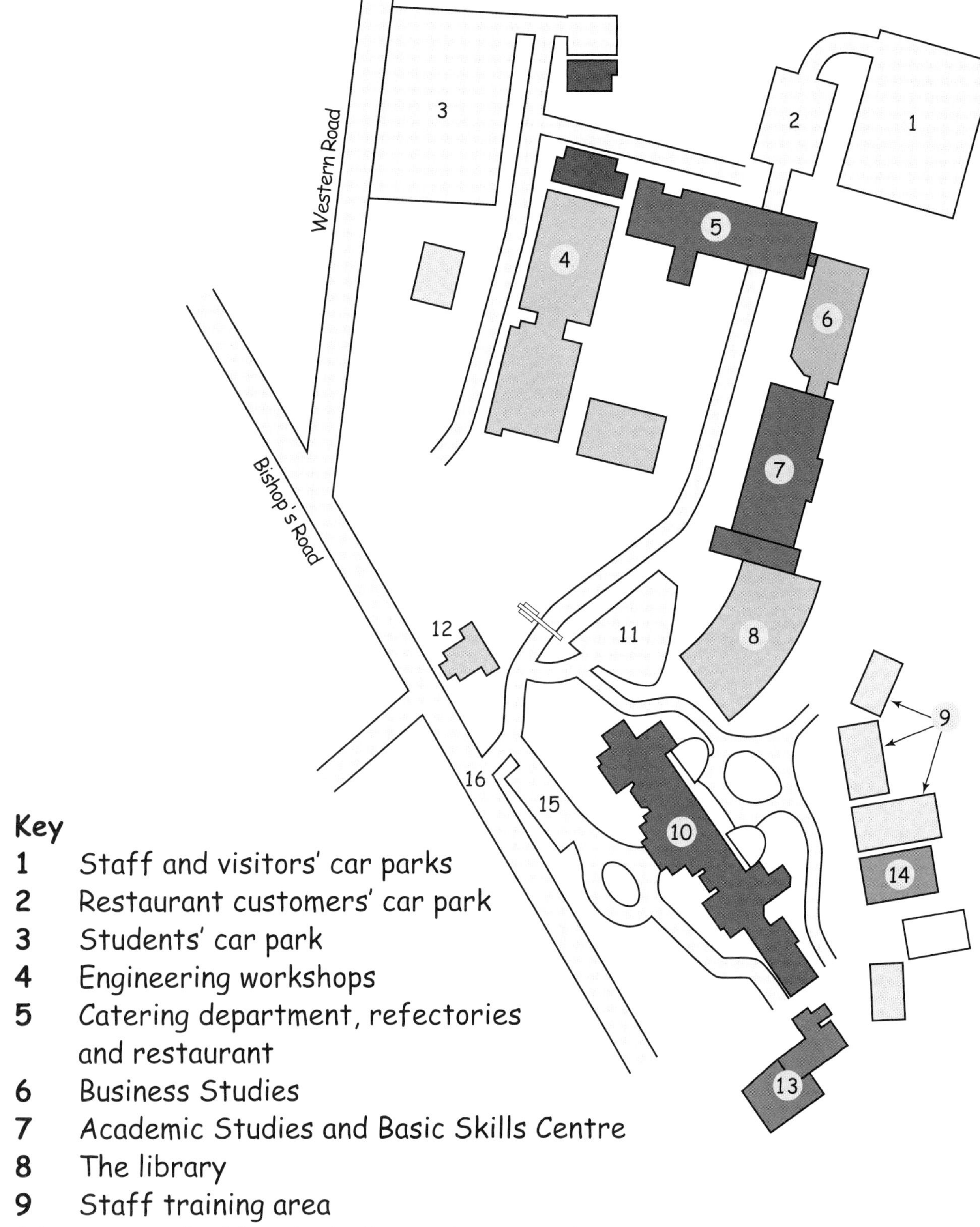


**Key**

**1** Staff and visitors' car parks
**2** Restaurant customers' car park
**3** Students' car park
**4** Engineering workshops
**5** Catering department, refectories and restaurant
**6** Business Studies
**7** Academic Studies and Basic Skills Centre
**8** The library
**9** Staff training area
**10** Yarrow Building including main reception
**11** Staff and visitors' car park
**12** Course enquiries and Admissions
**13** Students' Union and coffee shop
**14** Crèche and childcare services
**15** Staff and visitors' car park
**16** Main entrance

  

## College campus questions

HD1/E2.1

Use the map of the college campus to answer these questions.

What road is the main entrance to the college on?

______________________________________________

What is building 4?

______________________________________________

What number building is the crèche in?

______________________________________________

The catering department is in building 5. What else is in this building?

______________________________________________

How many buildings make up the staff training area?

______________________________________________

What road goes to the students' car park?

______________________________________________

Which building is the main reception in?

______________________________________________

What is in building 13?

______________________________________________

Name the buildings that are on either side of the Academic Studies and Basic Skills Centre.

______________________________________________

Which of the staff and visitors' car parks is nearest to the library?

______________________________________________

# Zoo map



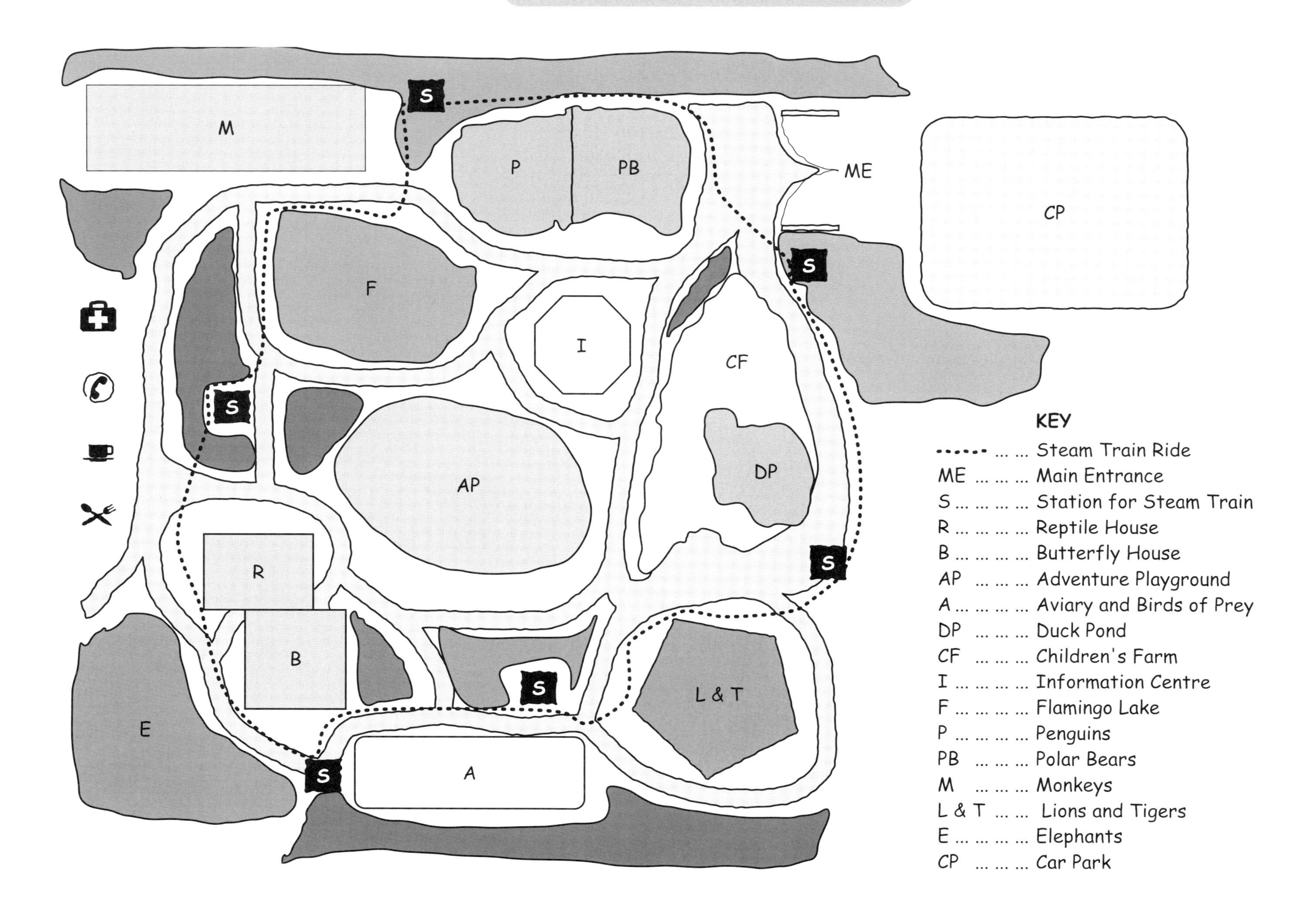

M
S
P
PB
ME
CP
F
I
CF
AP
DP
R
B
L & T
E
A
KEY
....... Steam Train Ride
ME ... ... Main Entrance
S ... ... ... Station for Steam Train
R ... ... ... Reptile House
B ... ... ... Butterfly House
AP ... ... Adventure Playground
A ... ... ... Aviary and Birds of Prey
DP ... ... Duck Pond
CF ... ... Children's Farm
I ... ... ... Information Centre
F ... ... ... Flamingo Lake
P ... ... ... Penguins
PB ... ... Polar Bears
M ... ... Monkeys
L & T ... Lions and Tigers
E ... ... ... Elephants
CP ... ... Car Park


  

## The zoo questions



Use the map of the zoo to answer these questions.

What is next to the polar bears?

______________________________________________

What other attraction is in the children's farm?

______________________________________________

What is next to the reptile house?

______________________________________________

Where are the elephants?

______________________________________________

Apart from walking, how can you get around the zoo?

______________________________________________

How many stations are there altogether?

______________________________________________

Are the monkeys near to a station?

______________________________________________



  

# Map of Broadstairs





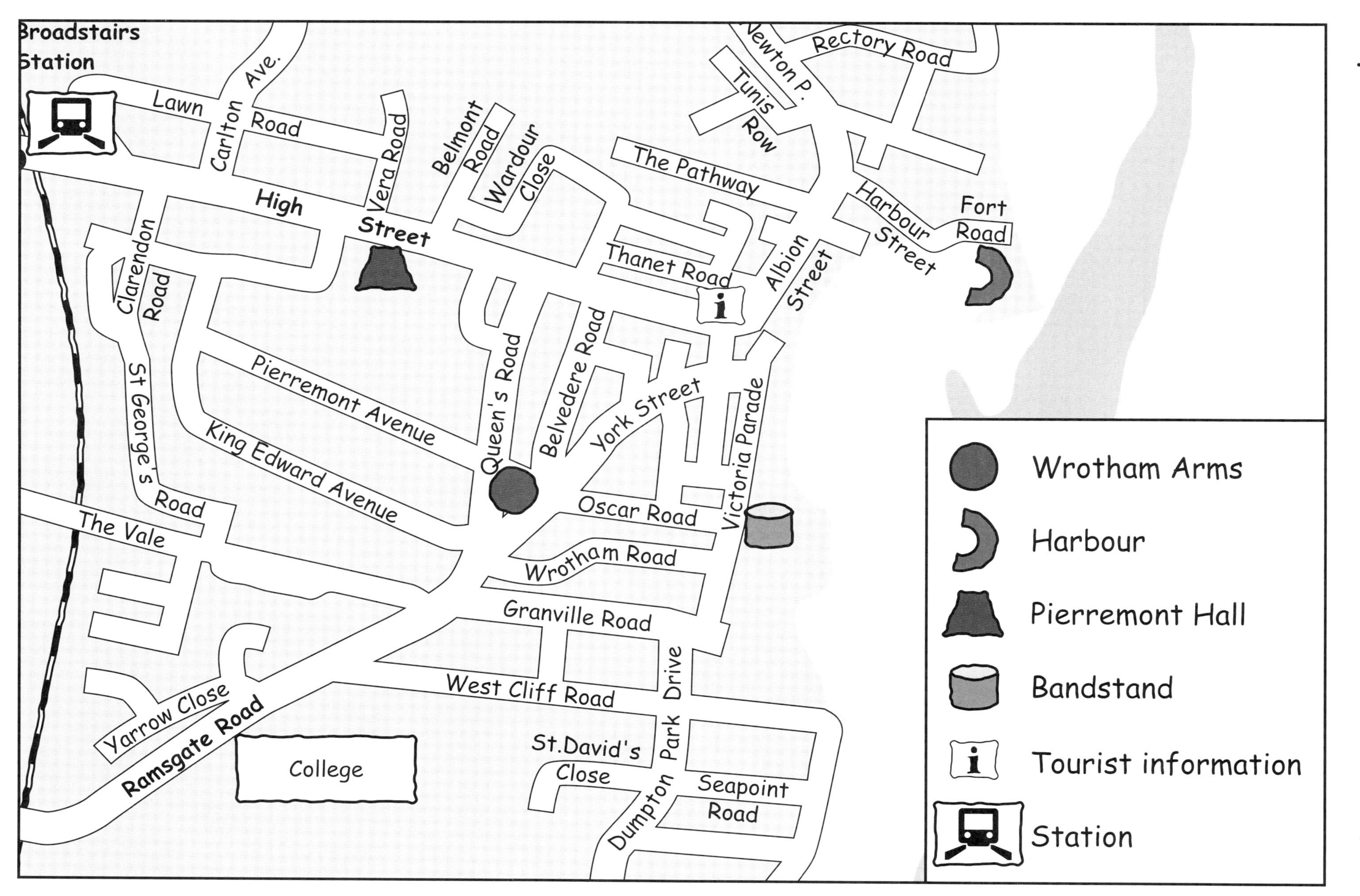




 

## Finding your way



Use the map of Broadstairs to answer these questions.

1 Which road is the college on? ______________________________

2 Which street is Pierremont Hall on? ______________________________

3 Is Fort Road near the harbour? ______________________________

4 Is the bandstand near Oscar Road? ______________________________

5 You walk from St David's Close to Seapoint Road.

Which road do you cross? ______________________________

6 Which road does Yarrow Close join? ______________________________

7 Which roads are nearest the Wrotham Arms?

______________________________ ______________________________

______________________________ ______________________________

8 Give directions from the station to the Tourist Information.

______________________________________________________________

______________________________________________________________



  

# The tube map



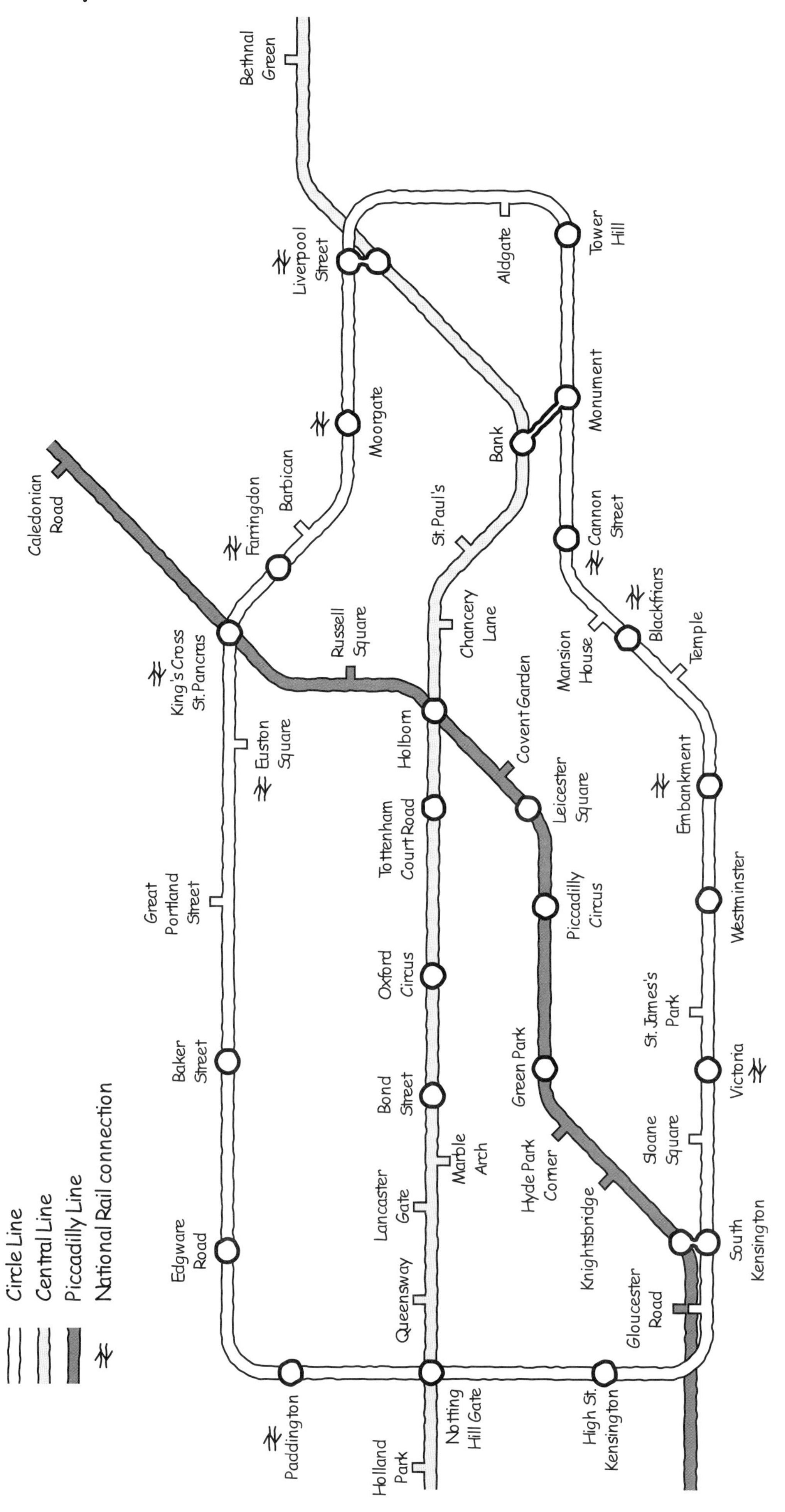
KEY:
Circle Line
Central Line
Piccadilly Line
National Rail connection
Bethnal Green
Liverpool Street
Aldgate
Tower Hill
Moorgate
Monument
Bank
Barbican
Farringdon
Caledonian Road
St. Paul's
Cannon Street
King's Cross St. Pancras
Russell Square
Chancery Lane
Mansion House
Blackfriars
Temple
Euston Square
Holborn
Covent Garden
Leicester Square
Embankment
Tottenham Court Road
Great Portland Street
Piccadilly Circus
Westminster
Oxford Circus
St. James's Park
Baker Street
Bond Street
Green Park
Victoria
Marble Arch
Hyde Park Corner
Sloane Square
Lancaster Gate
Knightsbridge
Edgware Road
South Kensington
Queensway
Gloucester Road
Paddington
Notting Hill Gate
High St. Kensington
Holland Park



 

## Catching the tube



Use the map of the tube to answer these questions.

1 How many stations are there on the Circle Line? ____________

2 How many stations on the Circle Line have National Rail connections?

____________

3 On the Circle Line how many stations are there **between** High St Kensington and Blackfriars going:

a via Victoria? ____________

b via King's Cross ____________

4 You want to go from High St Kensington to Marble Arch.

a Which lines will you travel on? ____________________________

b Where will you change trains? ____________________________

5 You want to go from King's Cross St Pancras to South Kensington. Which line has the least stops, the Piccadilly Line or the Circle Line?

____________________________________________________

6 On this map what are the stations at the ends of the Central Line?

____________________________________________________

7 How many stations are there **between** Caledonian Rd and Knightsbridge?

____________



  

## Sports

The diagram below is called a Venn diagram.

Each circle shows a group of people who do a sport.
Where the circles overlap a person does more than one sport.

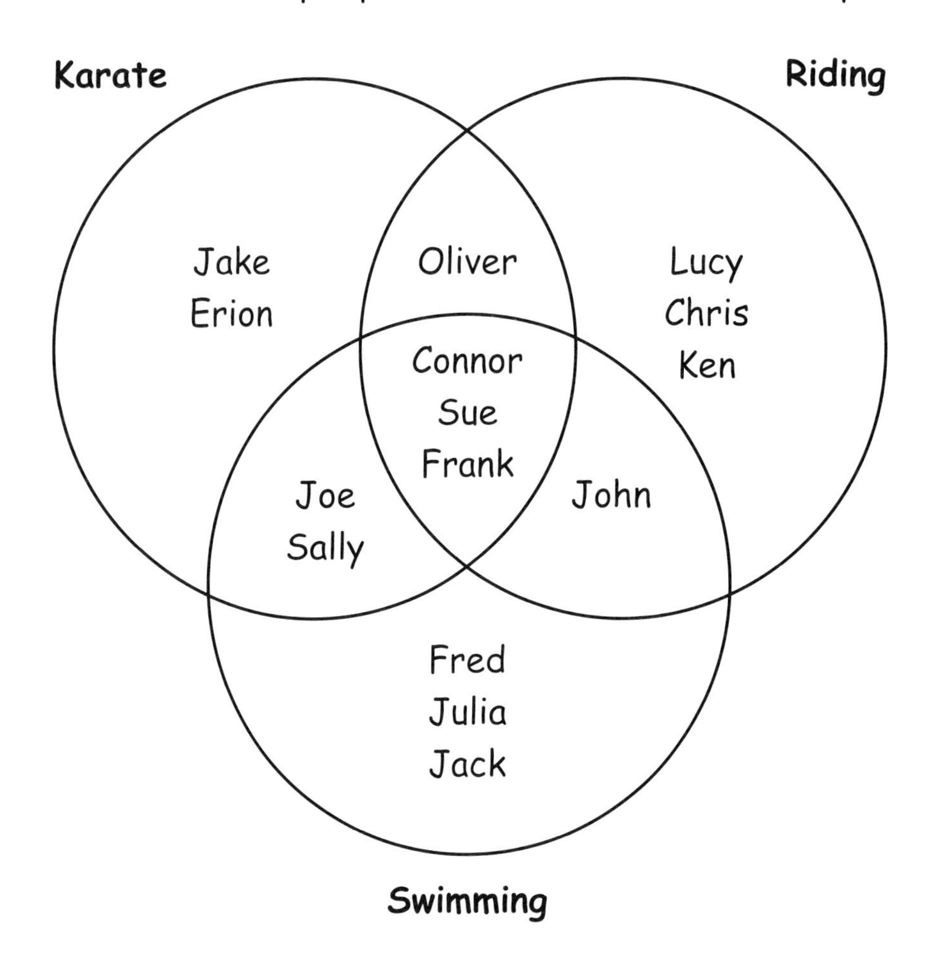




What sport does Erion do? ________________________________

Sally does karate and swimming.
What sports does John do? ________________________________

Who does all three sports? ________________________________

Who just swims? ________________________________

Who swims and does karate, but doesn't ride? ________________

 

## Pictograms

HD1/E1.1 (Preparation for HD1/E1.3)

A **pictogram** uses symbols to give information.

**Example**

Jane asks her friends what they like best - crisps, chocolate, biscuits, sweets or cakes. She draws a pictogram to show her results.

The **title** shows what it's about.

The **key** shows what each symbol means.

**Favourite treats**

**Key** 🯅 = 1 person

| | |
|---|---|
| **Crisps** | 🯅 🯅 🯅 🯅 🯅 🯅 |
| **Chocolate** | 🯅 🯅 🯅 🯅 🯅 🯅 🯅 🯅 🯅 |
| **Biscuits** | 🯅 🯅 🯅 🯅 🯅 |
| **Sweets** | 🯅 🯅 🯅 🯅 |
| **Cakes** | 🯅 🯅 🯅 🯅 🯅 🯅 🯅 |

The number of people who like crisps best is 6.

How many people like chocolate best? ____________________

How many people like biscuits best? ____________________

How many people like sweets best? ____________________

How many people like cakes best? ____________________

**Worksheets**

  

## Car sales



Denton Garage has sold the following cars this week:

| | | | | | |
|---|---|---|---|---|---|
| Ford | 5 cars | Vauxhall | 7 cars | Citroen | 9 cars |
| Peugeot | 4 cars | Rover | 10 cars | | |

Complete the pictogram below.
Remember to include a title.

**Key** 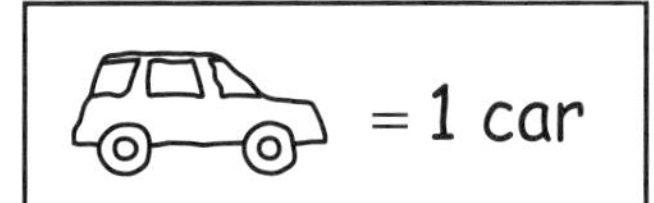


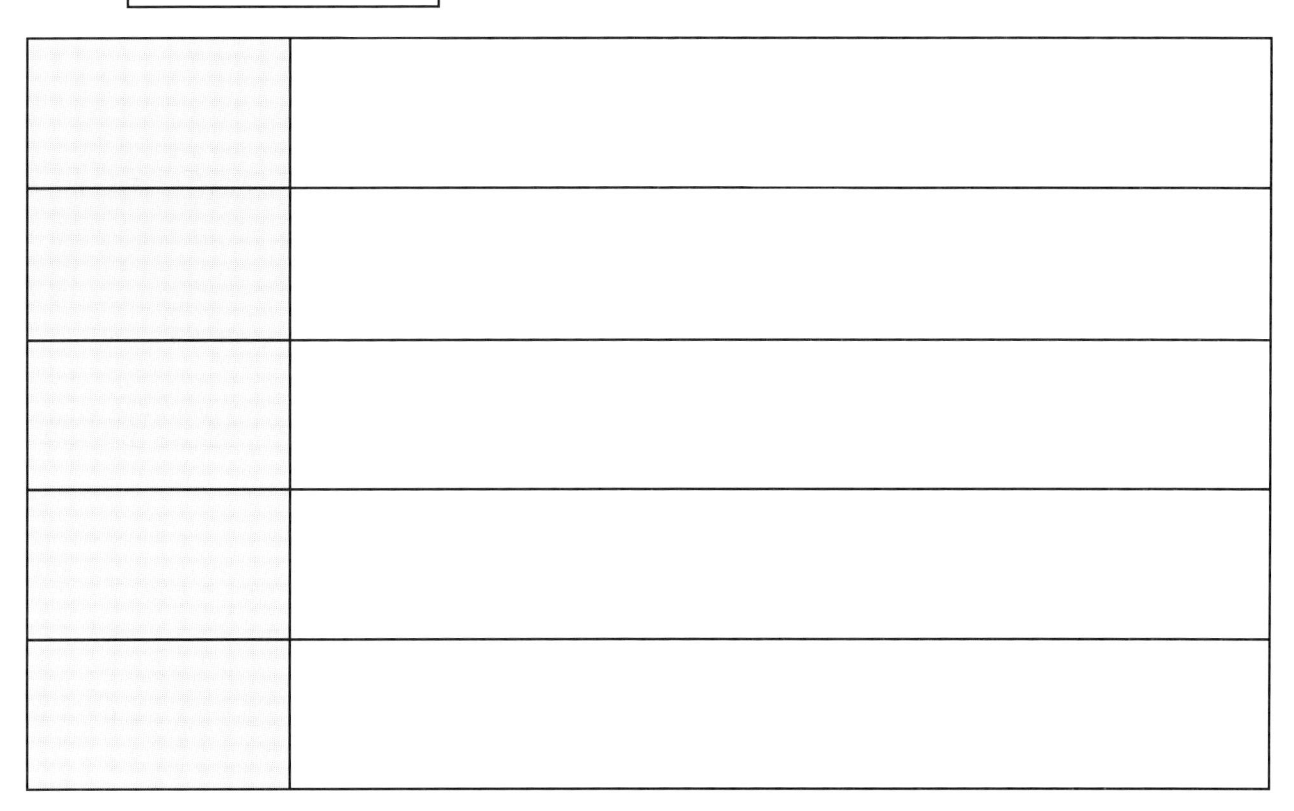

Which car sold the most? ______________________

How many Vauxhalls were sold? ______________________

How many Citroens were sold? ______________________

How many Rovers were sold? ______________________

Which car sold the least? ______________________

Which type of car sold 5? ______________________



 

## Saturday night

HD1/E1.3

A student asked the people in a class what they did on Saturday night. The results were:

| | | | |
|---|---|---|---|
| Went to cinema | 3 people | Met friends | 7 people |
| Watched TV | 9 people | Played music | 4 people |
| | | Read a book | 2 people |

Complete the pictogram below. Remember to include a title and key.

| | |
|---|---|
| **Cinema** | |
| **Met friends** | |
| **TV** | |
| **Music** | |
| **Book** | |

How many people went to a cinema? ______________________________

How many people met friends? ______________________________

How many people played music? ______________________________

What was the most popular? ______________________________

Which was least popular? ______________________________

How many people were asked? ______________________________



 

## Bed sales over 1 month at Sleepy Time Stores

Key = 5 beds

| | |
|---|---|
| King size bed | |
| Queen size bed | |
| Double bed | |
| Single bed | |
| Bunk beds | |

Which bed sold the most? ____________________

Which bed sold the least? ____________________

How many beds does a symbol represent? ____________________

How many king size beds were sold? ____________________

How many queen size beds were sold? ____________________

How many double beds were sold? ____________________

How many single beds were sold? ____________________

How many bunk beds were sold? ____________________

 

## Musical taste survey

Key  = 2 people

| | |
|---|---|
| **Blues** | 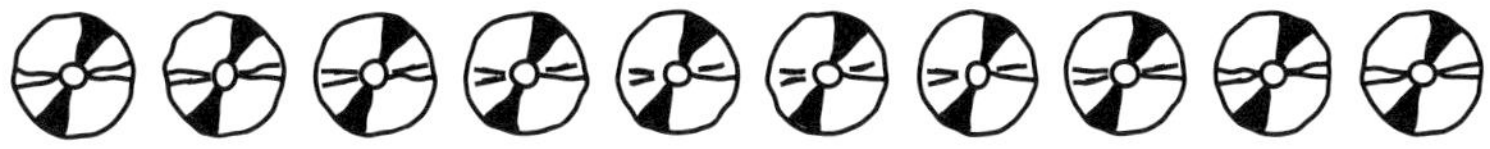 |
| **Heavy metal** | 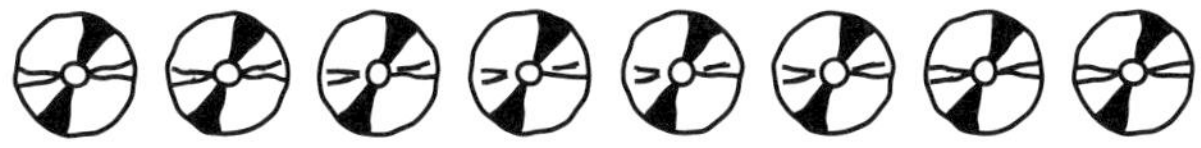 |
| **Pop** |  |
| **Country and western** |  |
| **Classical** |  |

What does  represent? ______________________

Which type of music is most popular? ______________________

Which type of music is least popular? ______________________

How many people like classical music? ______________________

How many people like blues? ______________________

How many people like heavy metal? ______________________

Which types of music are equally popular? ______________________

and ______________________

How many people took part in the survey? ______________________



  

## Getting to work



Below is a **bar chart**.

All charts need **labels** and a **title** to explain what they are about.

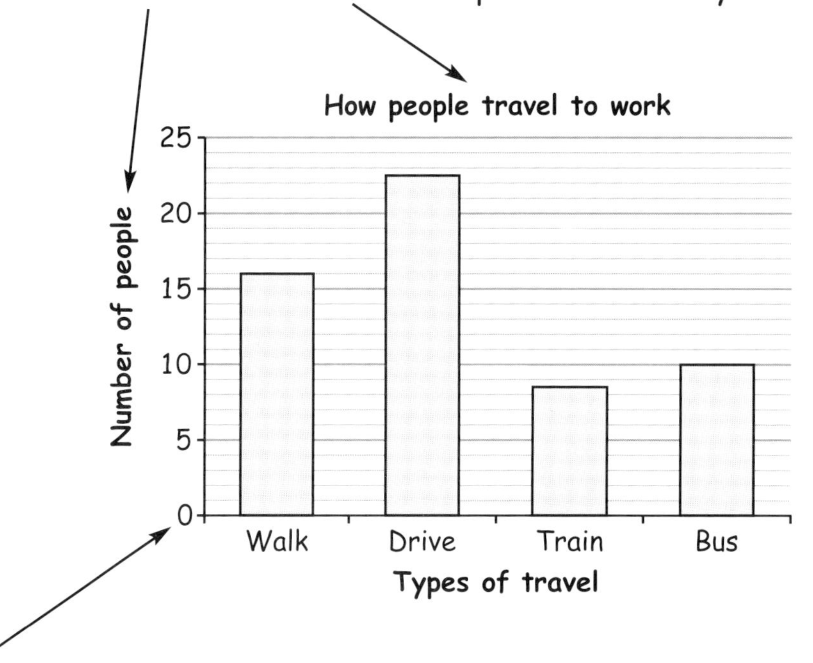


The numbers at the side are called the scale.

The scale shows how many people the bars represent.

What is the title of this chart? ______________________________

What is the most popular way of getting to work? ________________

What do the numbers on the scale go up in? ____________________

Complete this table.

| Type of travel | Number of people |
|---|---|
| Walk | |
| Drive | |
| Train | |
| Bus | |
| Total | |



  

# Temperature

HD1/E2.1, HD1/E2.2

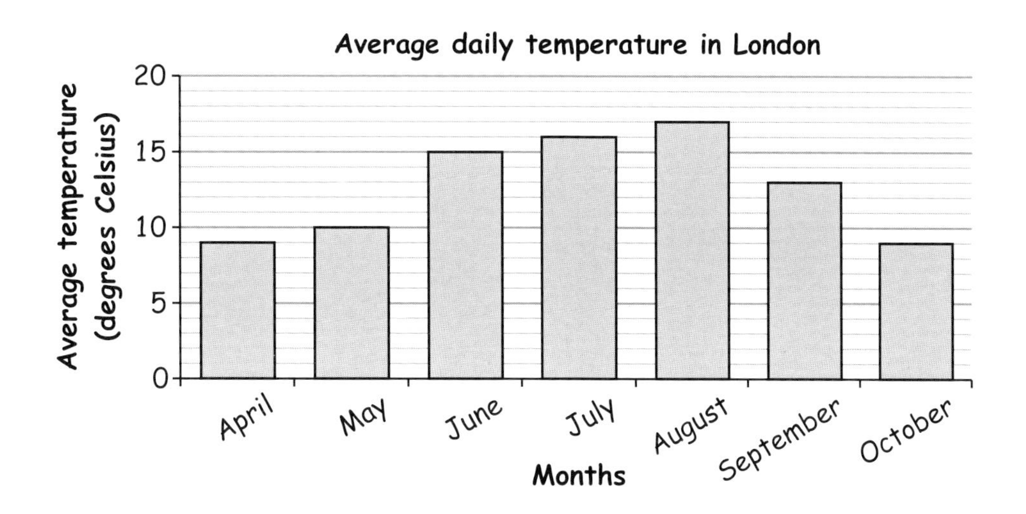


Use the chart to answer the questions.

What does this chart show? ______________________________

What do the numbers at the side represent? ______________________

What does the scale go up in? ______________________________

What was the average temperature in June? ______________________

Which month was the hottest? ______________________________

Which month was the coldest? ______________________________

How much warmer was it in September than October? _____________

How much colder was it in April than June? _____________

Worksheets

## Cat food sales

HD1/E2.1, HD1/E2.2

This chart shows the amount spent on cat food at a supermarket in a day.

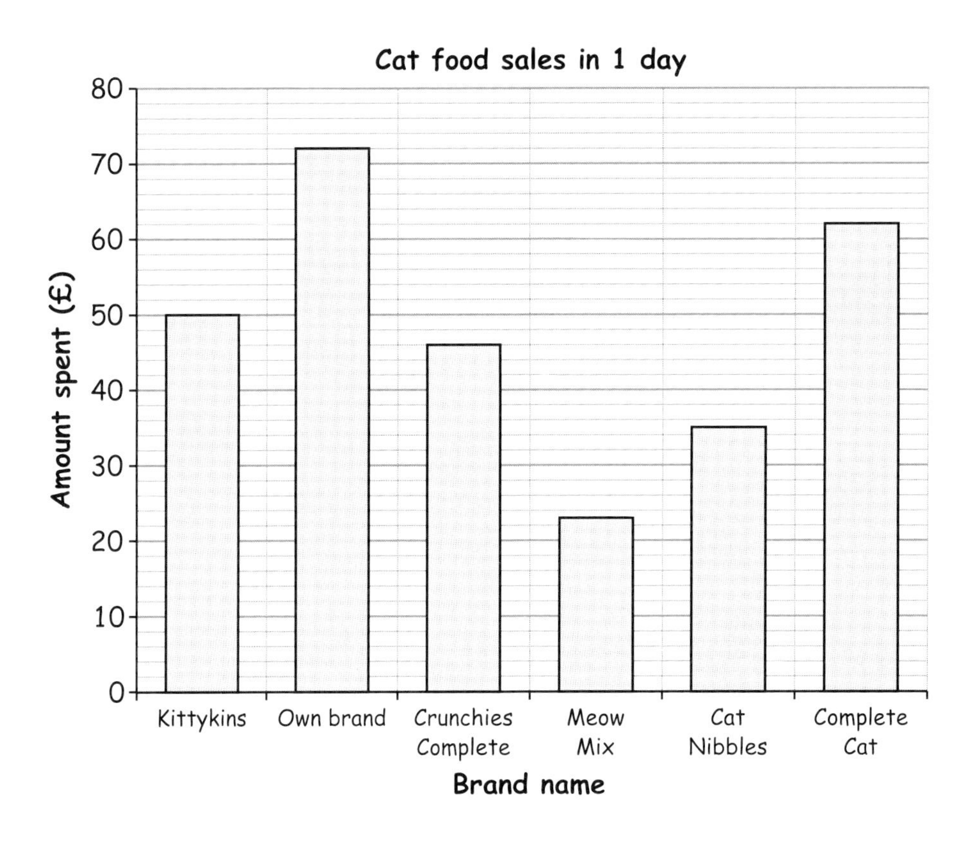


What does each division on the scale represent? ____________________

How much was spent on Complete Cat? ____________________

Which cat food was most spent on? ____________________

How much was spent on the most popular cat food? ____________________

List the brands in order of how much was spent on them.
Start with the most popular brand.

______________________________________________________________

______________________________________________________________

Worksheets

 

## Game shows

HD1/E2.1, HD1/E2.2

Use the bar chart to answer the questions below.

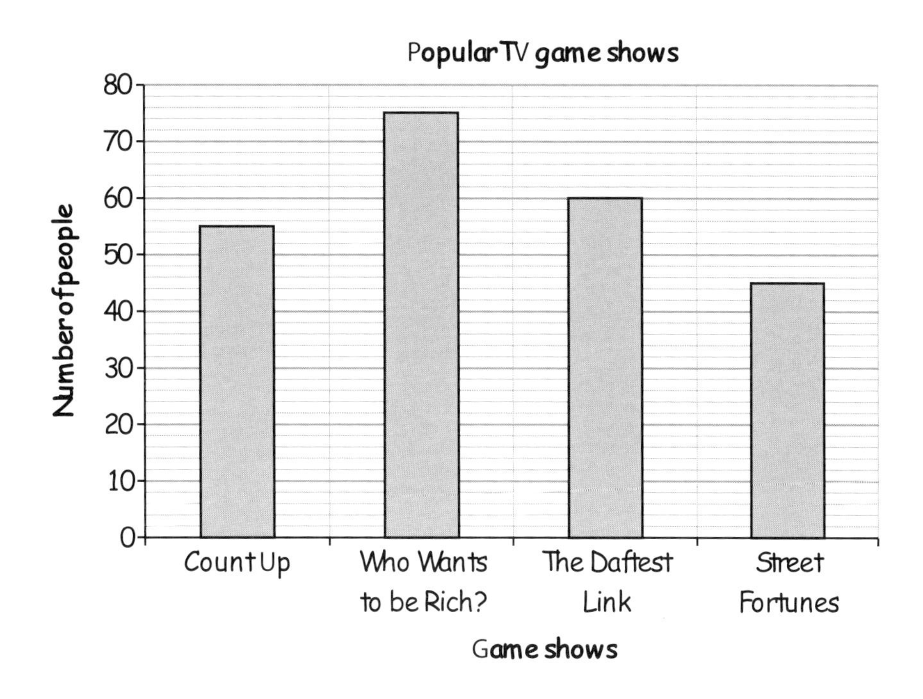


What is the bar chart about? ______________________________

Which is the most popular game show? ______________________

How many people chose it? ________________________________

Which is the least popular game show? ______________________

How many people chose it? ________________________________

How many more people chose The Daftest Link than Count Up?

__________________________________________________________

Worksheets

  

# First drink of the day

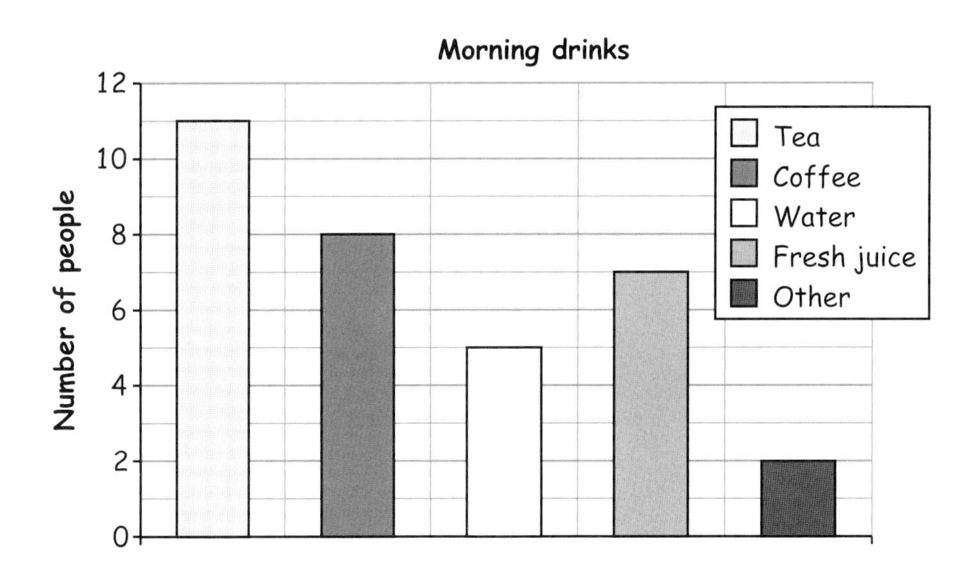




What is this bar chart about? ___________________________

How many people drink coffee? ___________________________

How many people drink water? ___________________________

Which is the most popular drink? ___________________________

How do you know this? ___________________________

How many people drink tea? ___________________________

How many people took part in the survey? ___________________________

Why do you think there is a box marked other? ___________________________

___________________________

 

# Speed trap

HD1/E2.1, HD1/E2.2

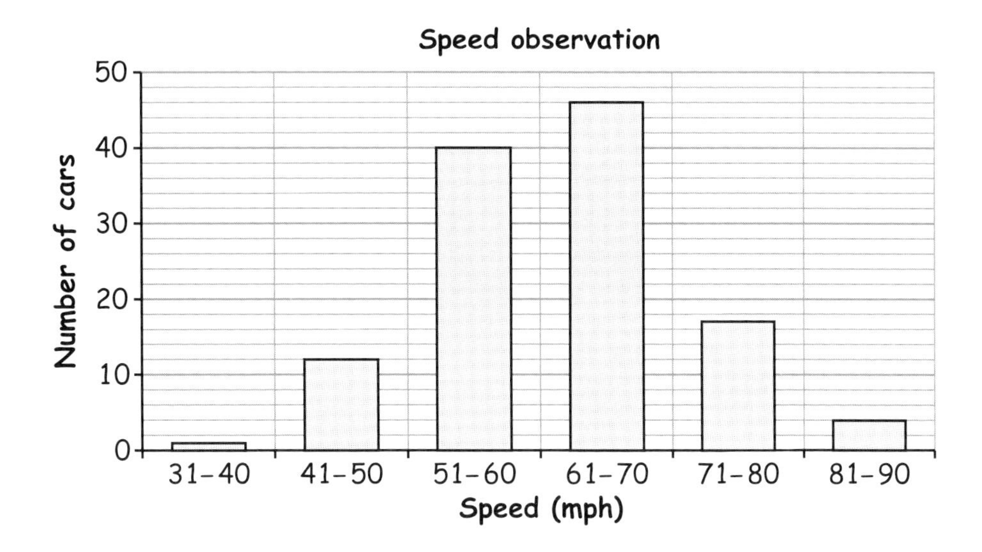


What speeds did the most cars travel between? ________________

Complete this table.

| Speed (mph) | Number of cars |
|---|---|
| 31-40 | |
| 41-50 | |
| 51-60 | |
| 61-70 | |
| 71-80 | |
| 81-90 | |

How many cars were travelling between 31 and 50 mph? ______________

How many cars were travelling between 51 and 70 mph? ______________

The speed limit on the road is 70 mph.

How many cars were breaking the speed limit? ____________________

How many cars were travelling at 60 mph or less? __________________

Worksheets

  

# Fish sales

HD1/E2.1, HD1/E2.2

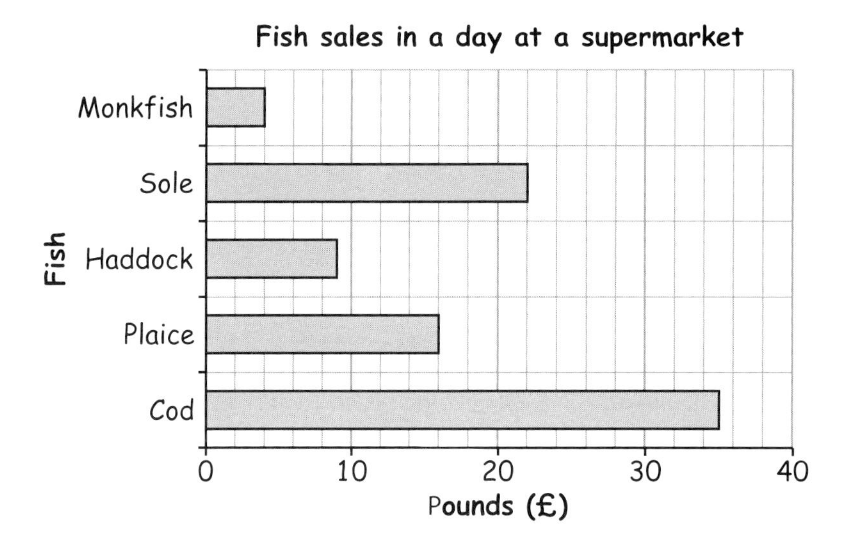


Worksheets

What does the chart show? ______________________________

Which fish sold the most? ______________________________

Which fish sold the least? ______________________________

What do the grid lines go up in? ______________________________

How much was spent on monkfish? ______________________________

How much was spent on haddock? ______________________________

Was more spent on plaice or sole? ______________________________

 

# Property sales

HD1/E2.1, HD1/E2.2

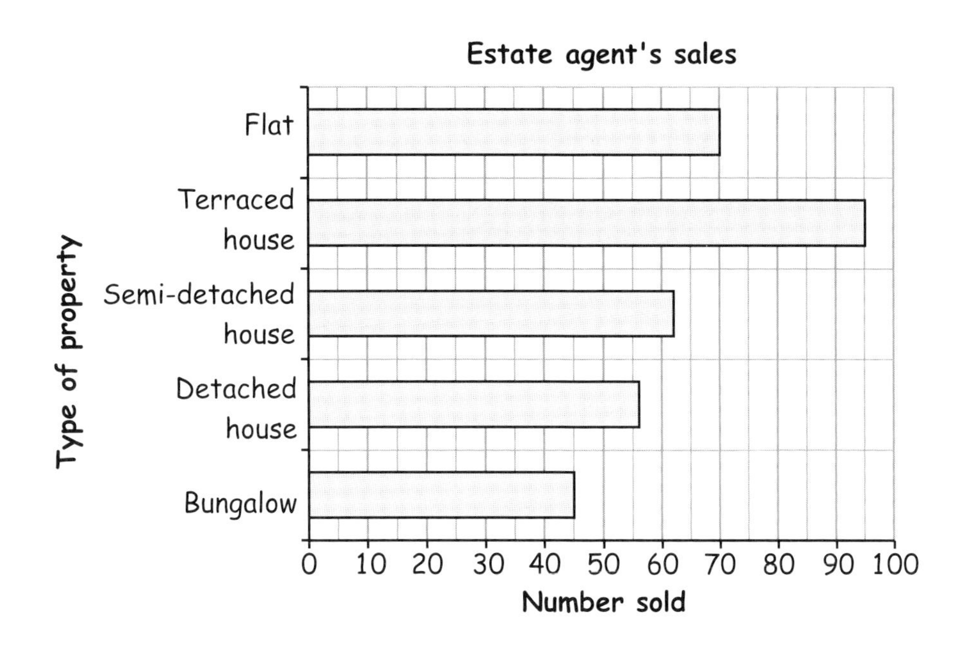


What is the bar chart about? ______________________________

What type of property sold most? ______________________________

What does the scale go up in? ______________________________

Which type of property sold 62? ______________________________

How many flats were sold? ______________________________

Which type of property sold the least? ______________________________

Why do you think that was? ______________________________

______________________________________________________________

Worksheets

  

## Leisure centre classes

HD1/E2.1, HD1/E2.2

This chart compares the number of men and women who go to classes at a leisure centre.

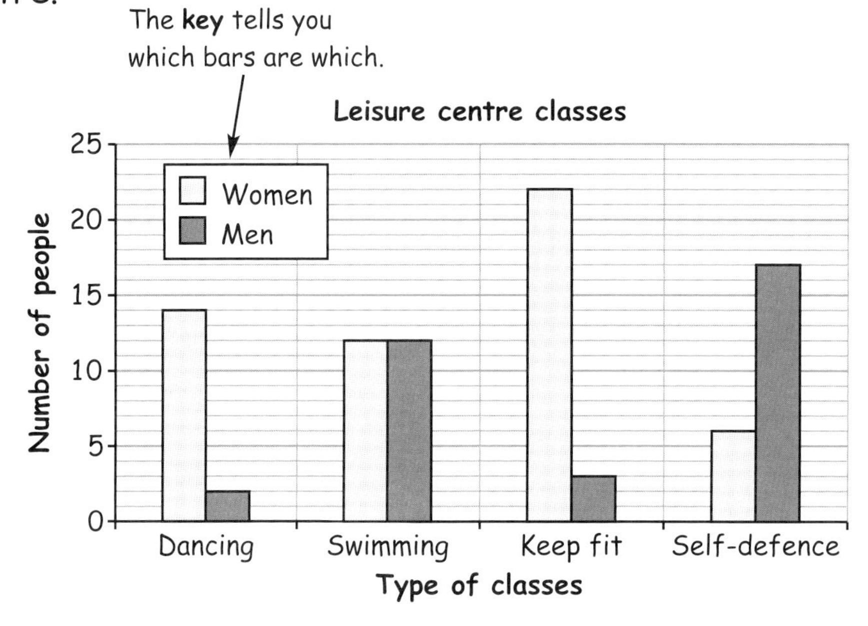


Which class has the same number of men and women? ________________

________________________________________

Which class has more men than women? ________________

________________________________________

Which class has the greatest number of women? ________________

________________________________________

Which class has the smallest number of men? ________________

________________________________________

Complete this table.

| **Class** | **Number of men** | **Number of women** | **Total** |
|---|---|---|---|
| Dancing | | | |
| Swimming | | | |
| Keep fit | | | |
| Self-defence | | | |
| **Total** | | | |

Worksheets

 

# Smellies

HD1/E2.1, HD1/E2.2

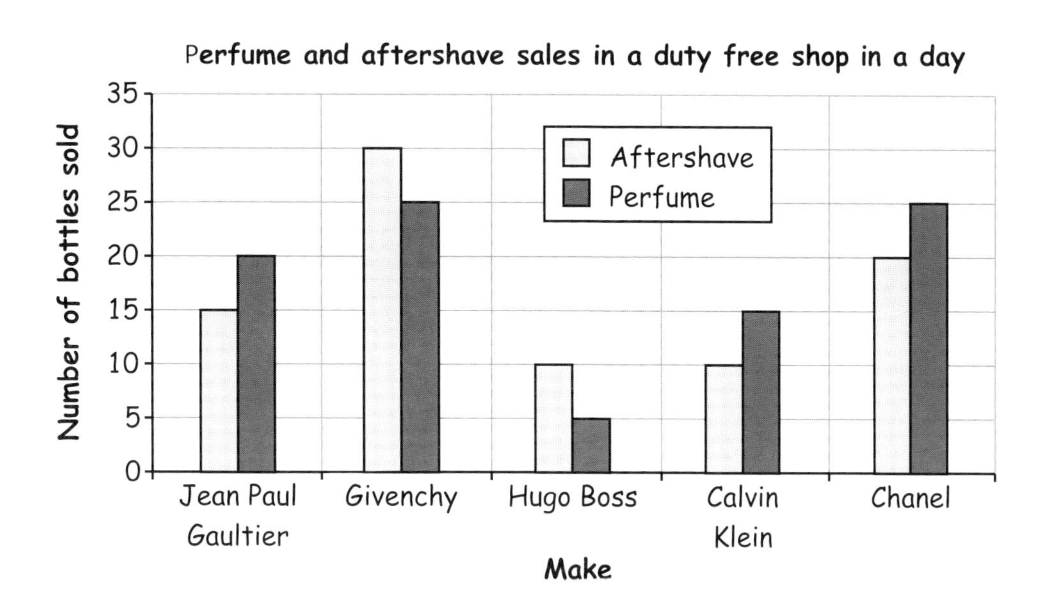


What does the chart show? ______________________________

Why are the bars in two different colours? ______________________________

Did John Paul Gaultier sell more aftershave or perfume? ______________

Which is the best selling aftershave? ______________________________

Which are the best selling perfumes? ______________________________

Is it true that Calvin Klein sold more aftershave than perfume? ________

How many bottles of perfume were sold altogether? ______________

How many bottles of aftershave were sold altogether? ______________

Which make sold the least? ______________________________

Worksheets

 

# What is missing? A

HD1/E2.1, HD1/E2.2

Only one of the four charts below is correct.

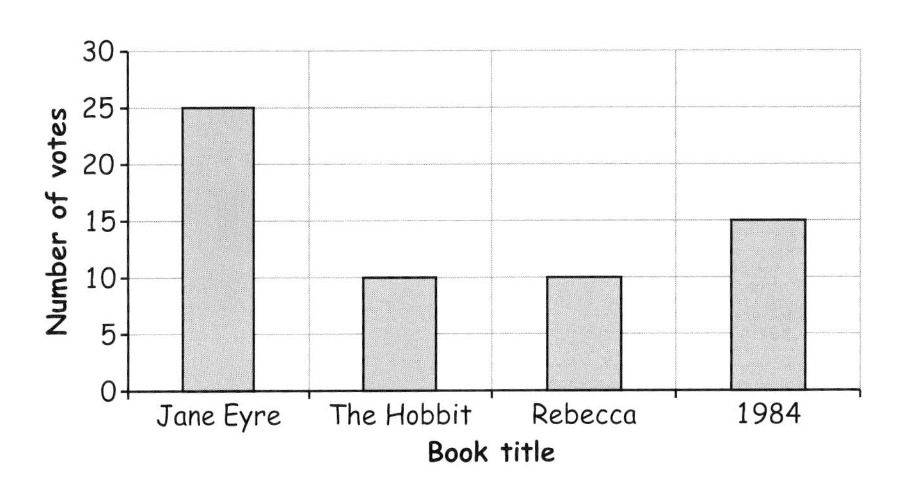


Is there anything missing from this chart? If so, what?

---

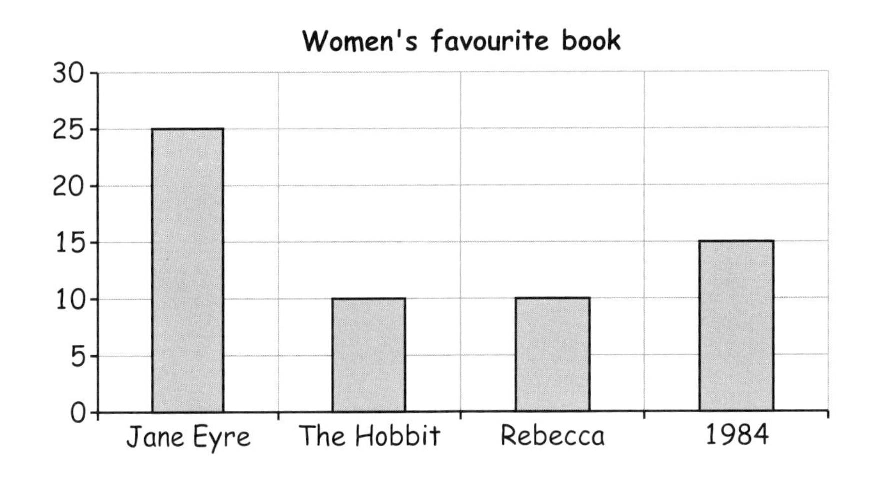


Is there anything missing from this chart? If so, what?

---

Worksheets

 

# What is missing? B

HD1/E2.1, HD1/E2.2

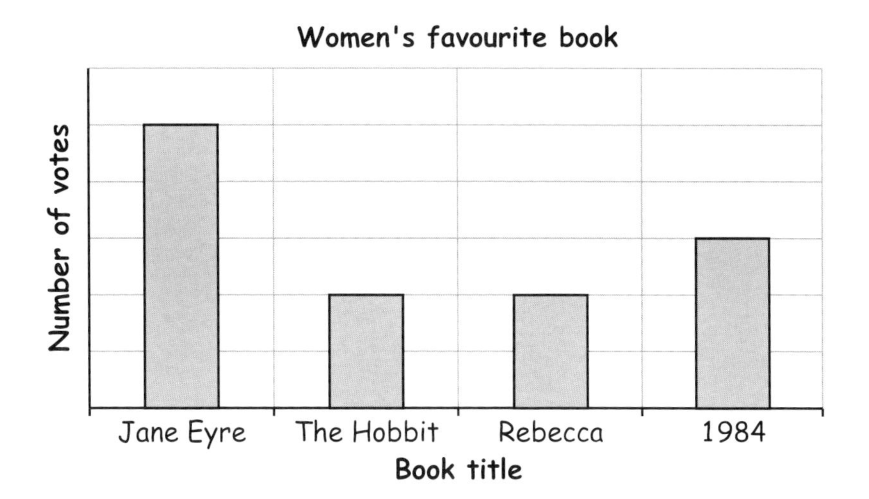


Is there anything missing from this chart? If so, what?

________________________________________

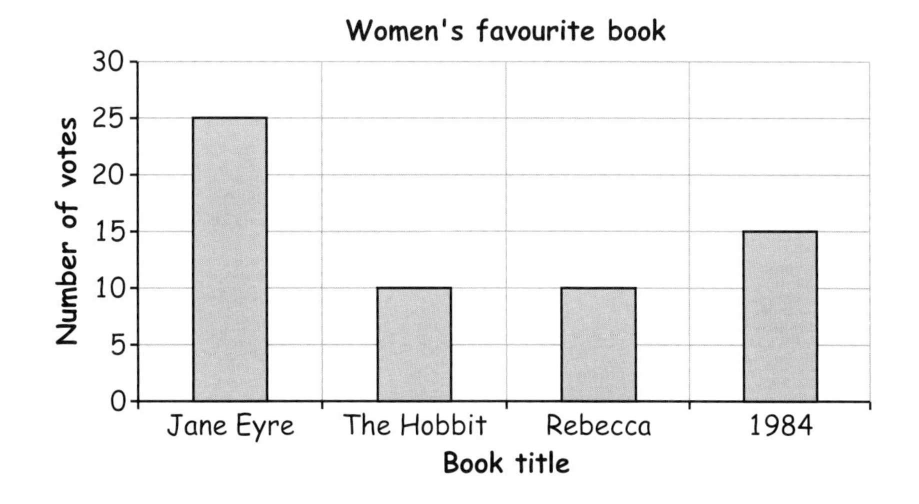


Is there anything missing from this chart? If so, what?

________________________________________

Worksheets

  

## Names

HD1/E1.2, HD1/E1.3

Below is a list of names.

Write them in 2 lists – a list of men's names and a list of women's names.

| | **Men's names** | **Women's names** |
|---|---|---|
| Julie | | |
| Brian | | |
| Samantha | | |
| Isobel | | |
| Raymond | | |
| Lisa | | |
| Nigel | | |
| Lucy | | |
| Jane | | |
| Philip | | |
| Joshua | | |
| Anne | | |
| Eric | | |
| Anthony | | |





 

## Sort the washing

HD1/E1.2, HD1/E1.3

Here is a list of clothes to be washed.

| | | | |
|---|---|---|---|
| Yellow towel | White towel | Red jumper | Black gloves |
| White shirts | Green top | Blue jeans | White socks |
| Red socks | Black trousers | White hankies | White T-shirt |
| White tablecloth | Red tie | White tea towel | Pink T-shirt |

The **white** items must be washed separately.
Make a list of the white items beside the washing basket.

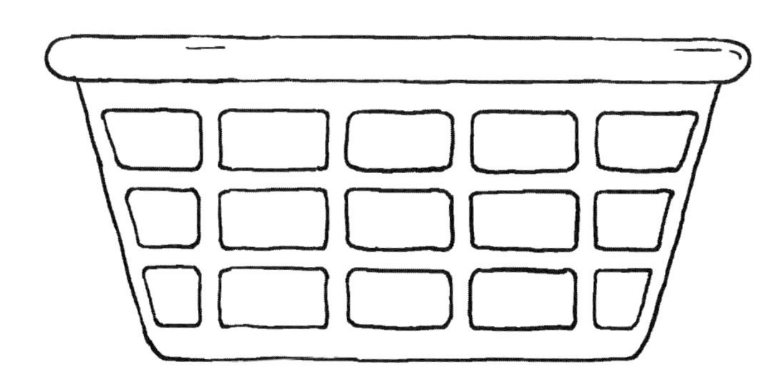

Worksheets

  

## Recycling

Drink cans go in the **drink cans** recycling bank.

All other tins go in the **tins** bank.

Draw an arrow from each tin and can to the correct recycling bank.

**Drink cans** **Tins**

How many tins go into the tins bank? ____________________

How many cans go into the drink cans bank? ____________________



 

## Department store



You work in the local department store.
You have to sort out a delivery.

Colour the items for the **baby** department **blue**.

Colour the items for the **shoe** department **red**.

What type of items are **not coloured**? ______________________________

__________________________________________________________________

## Cook a vegetarian meal

HD1/E1.2

You want to cook a meal for a vegetarian.

You look at these recipes for ideas.

Circle the recipes that are vegetarian.

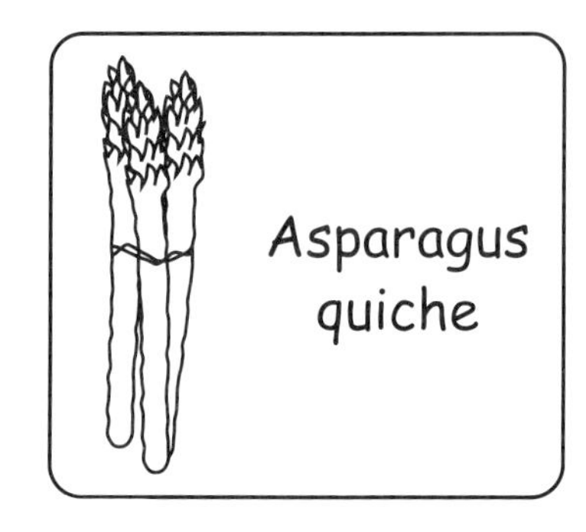

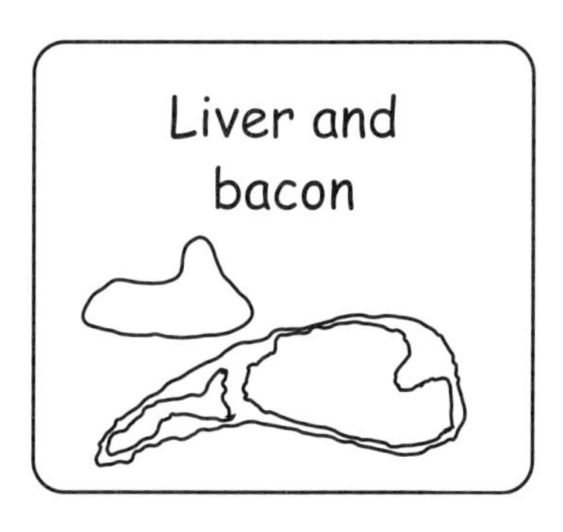


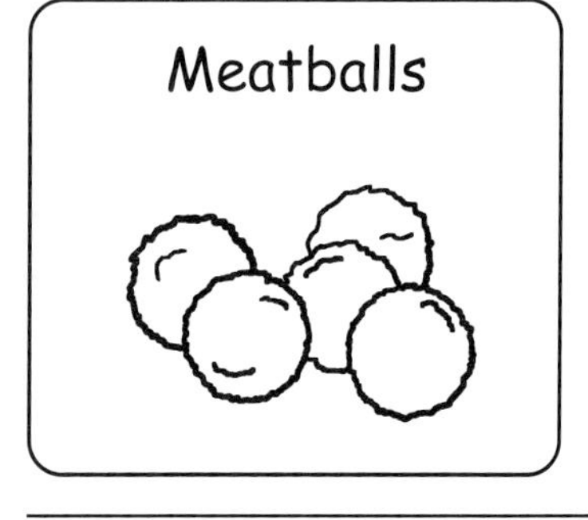

## Food shopping



You go shopping. The things you need are shown below.
Write a shopping list.

- Put all the items from the baker's first
- then all the items from the greengrocer's (i.e. the vegetable shop)
- then everything else at the end of the list.

Sausage rolls

Wrapping paper



  

## DIY tools



You are clearing out your garage.
The things you find are listed below.
You put all the do-it-yourself tools together.

Circle the items that are do-it-yourself tools.

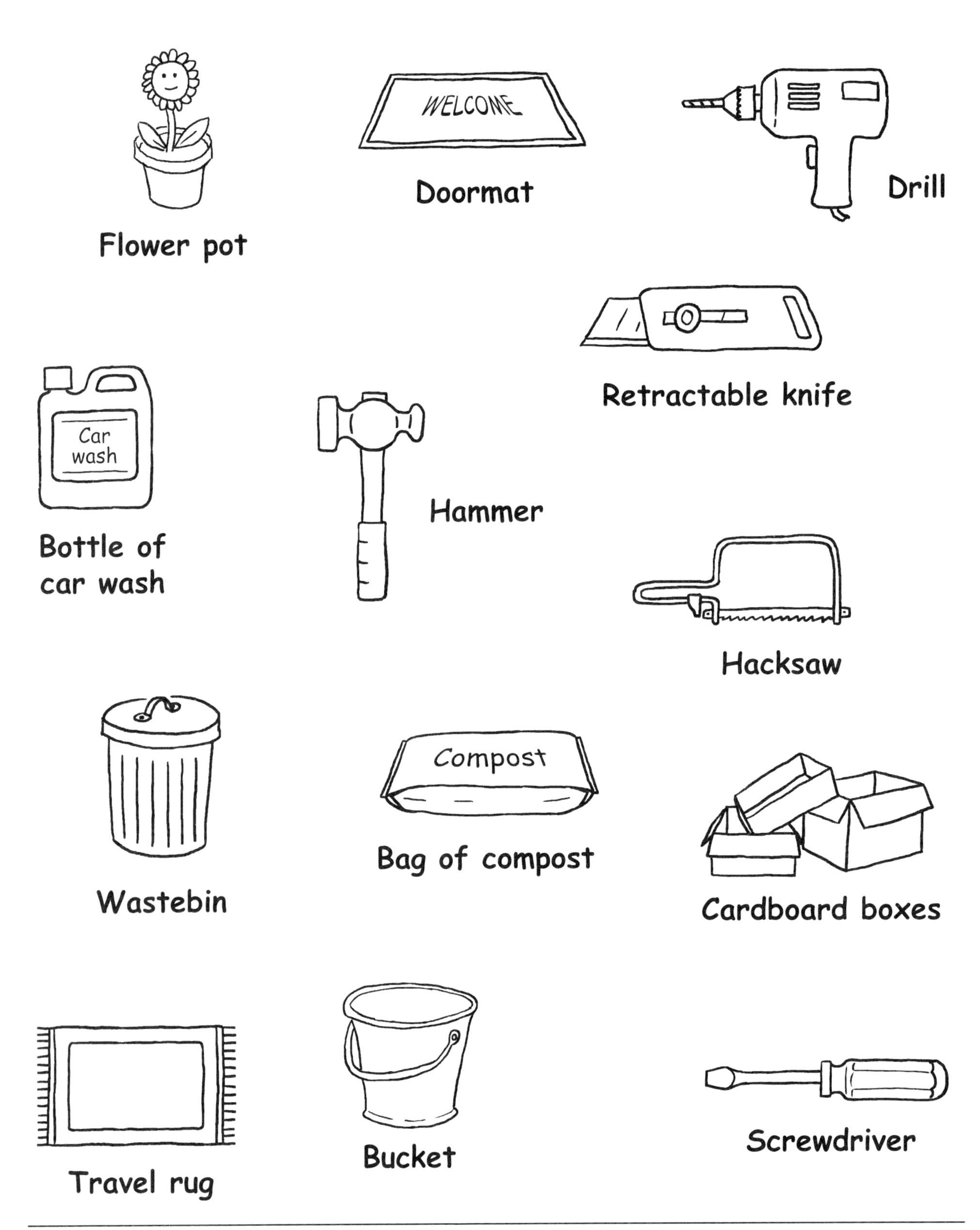


Flower pot

Doormat

Drill

Retractable knife

Bottle of car wash

Hammer

Hacksaw

Wastebin

Bag of compost

Cardboard boxes

Travel rug

Bucket

Screwdriver





 

## Shapes



Some of the shapes below are 2-D (flat).
Others are 3-D (not flat).

Colour the 2-D shapes in.

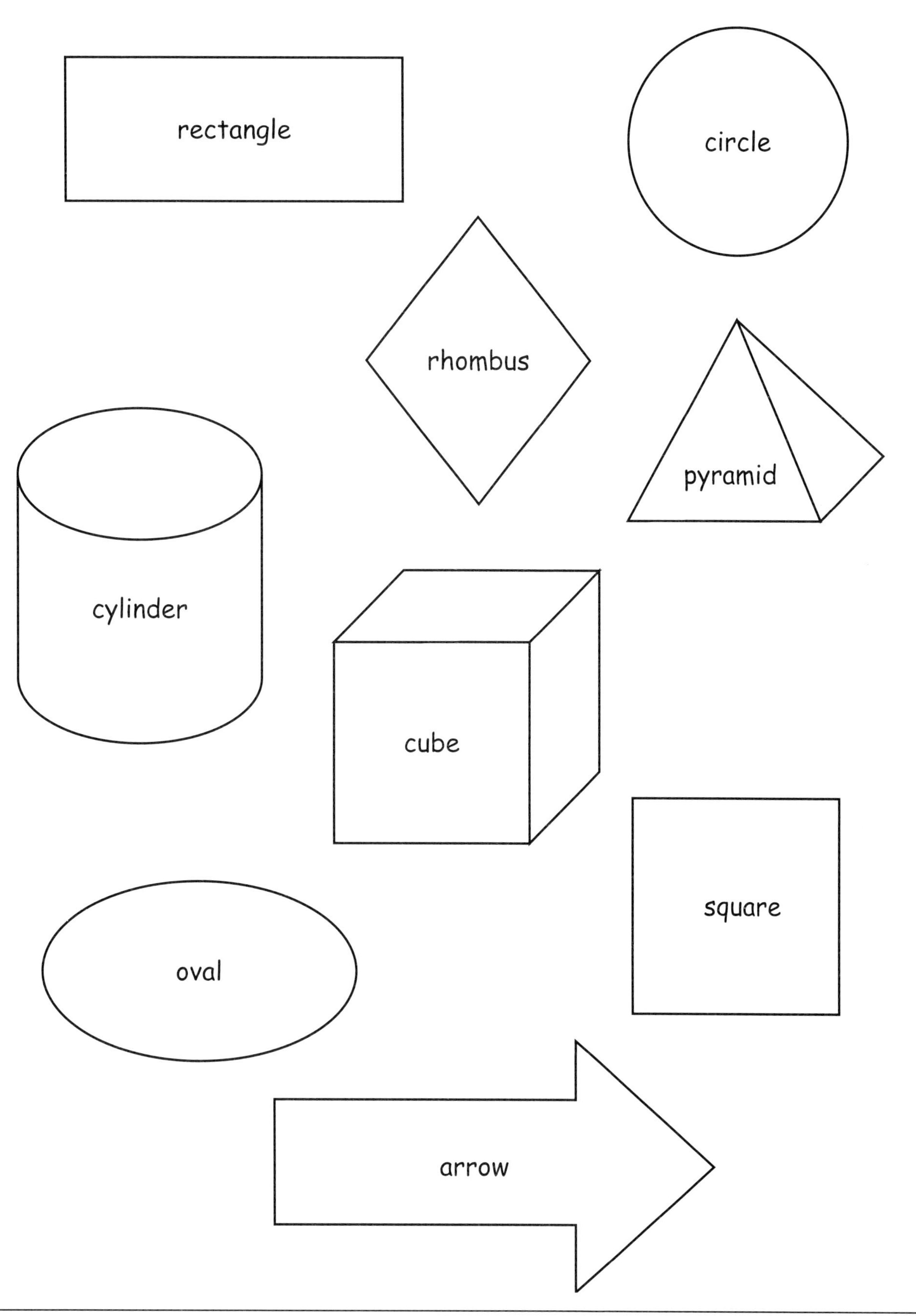

## Measuring things

All of these things can be measured, but not all with a tape measure.

Put a **tick** beside those you **can measure with a tape measure**.
Put a **cross** beside those you **can't measure with a tape measure**.

The temperature

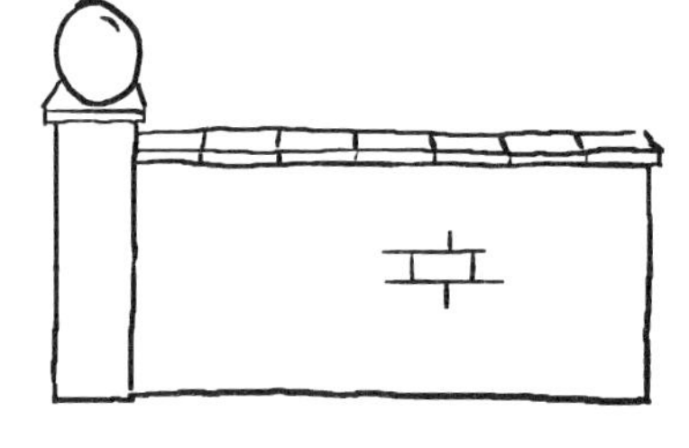

The height of a wall

How long it takes to run a race

The length of a rug

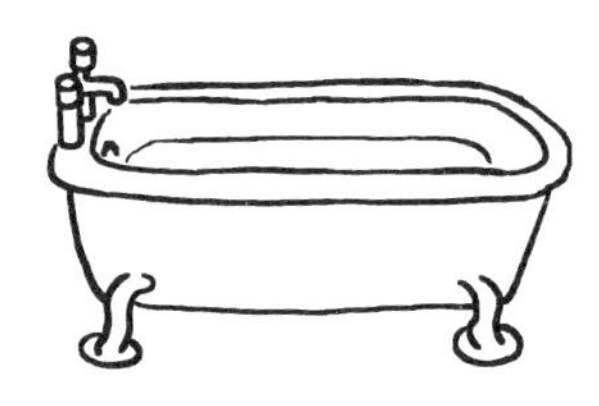

How hot the water is

A child's height

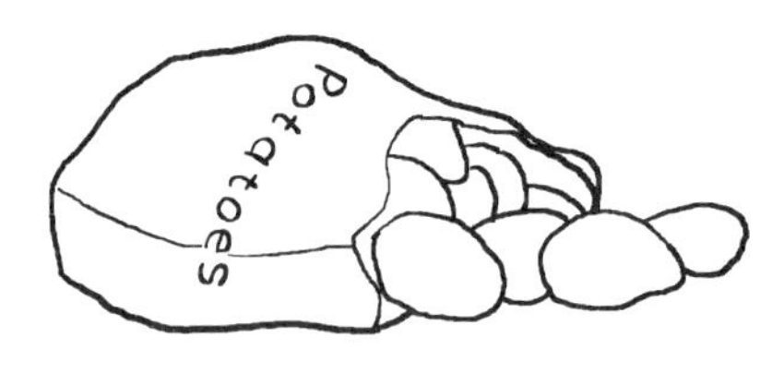


The weight of some potatoes

How much rain has fallen

The length of a skirt

How long a programme lasts



 

## Sort them

HD1/E1.2

These books have been sorted into 3 piles:

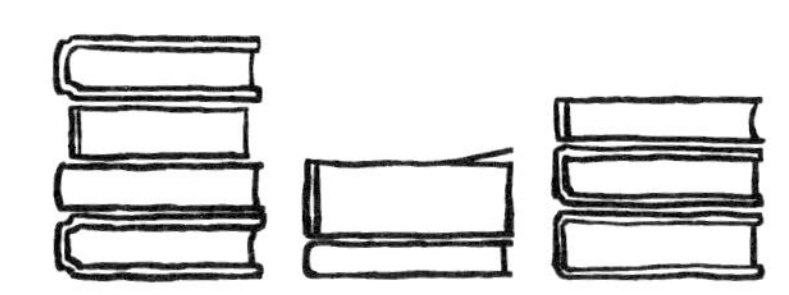

English books, maths books and science books.

How have they been sorted?

By: Size Author Subject ← Circle one of these

These films are on 3 different shelves at the video shop.

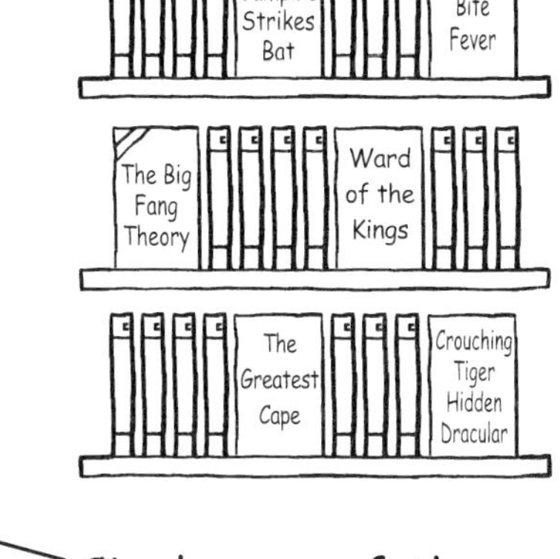


Universal 12 and above 18 plus

How have they been sorted?

By: Age Genre DVD or video ← Circle one of these

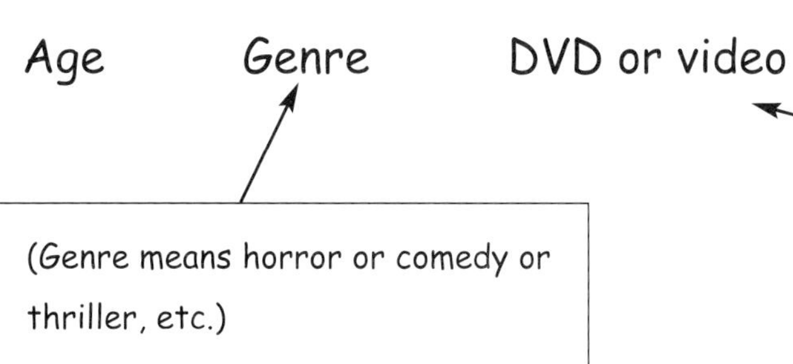

(Genre means horror or comedy or thriller, etc.)


The local chemist has put:

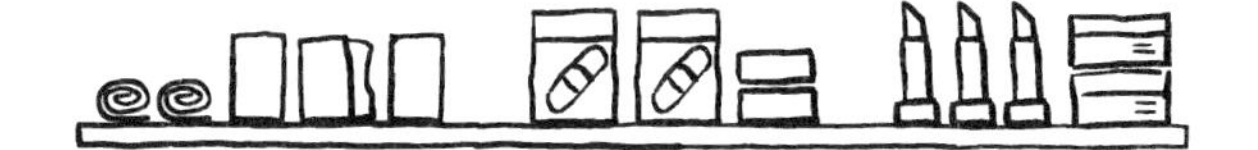

- the bandages and plasters together
- the pain killers together
- the make-up goods together.

How have they been sorted? Circle one of them

Price Type Size ← Circle one of these

Worksheets

  

## Football supporters

HD1/E2.3

A football club sells scarves and hats.

Draw a circle round the supporters who are wearing **both** a scarf **and** hat.
Draw a triangle round the supporter who are wearing a **scarf but no hat**.
Draw a square round the supporter who are wearing a **hat but no scarf**.

How many supporters are wearing **both** a scarf **and** hat? ___________

How many supporters are wearing a **scarf but no hat**? ___________

How many supporters are wearing a **hat but no scarf**? ___________

Worksheets

 

## On the hour

HD1/E2.3

The clocks all tell different times.

Circle the clocks that are digital **and** on the hour.

Worksheets

# Jumbled shapes



Below are lots of different shapes.

Colour shapes with straight edges and 4 sides blue.

Colour shapes with straight edges and more than 4 sides red.

Colour shapes with no straight edges and only one side green.

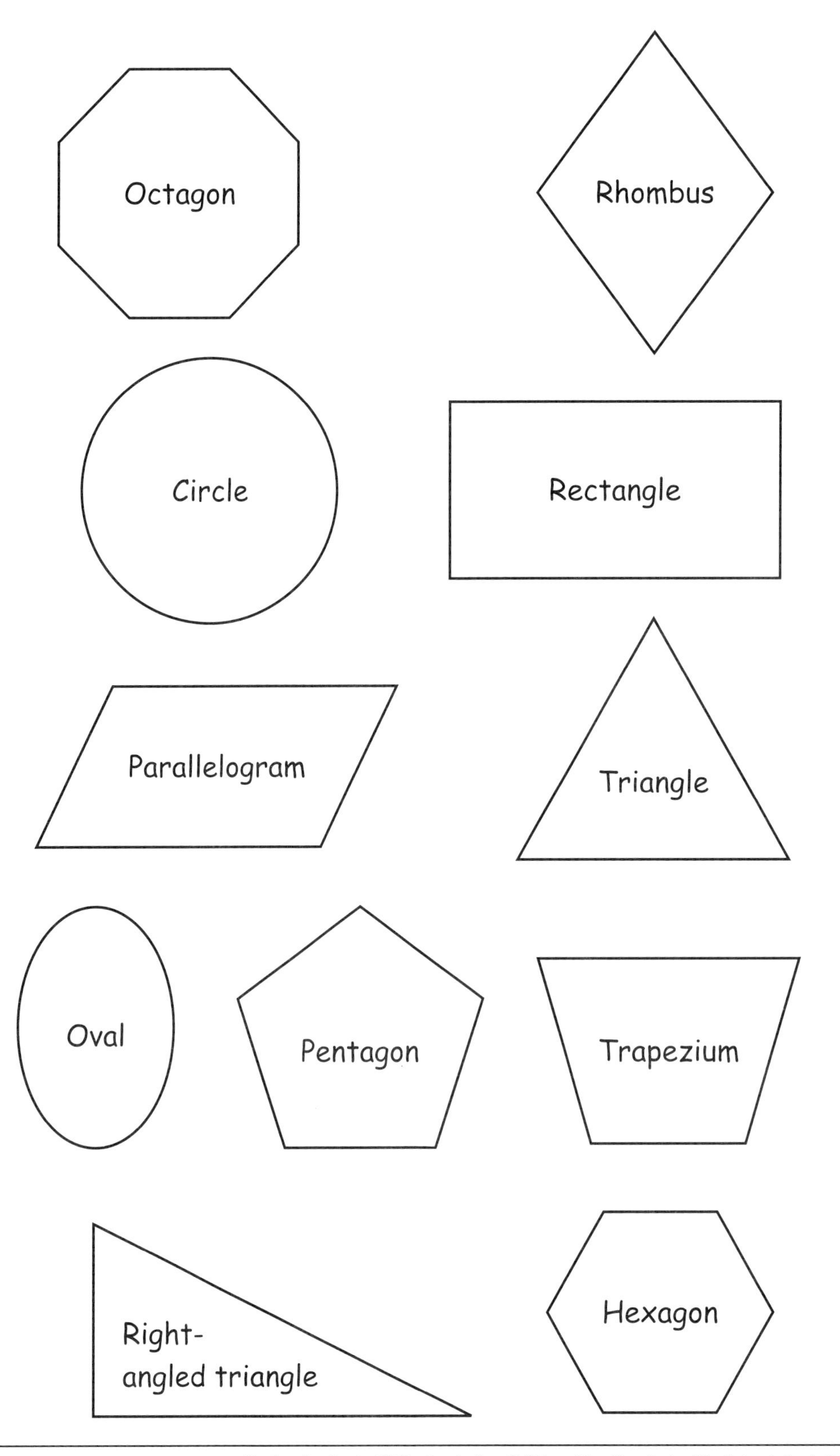




 

## Breakfast

HD1/E2.3

A café records what people eat for breakfast.

How many people have beans **and** tomatoes? ____________________

How many people have sausage **and** beans? ____________________

How many people have sausage **and** egg? ____________________

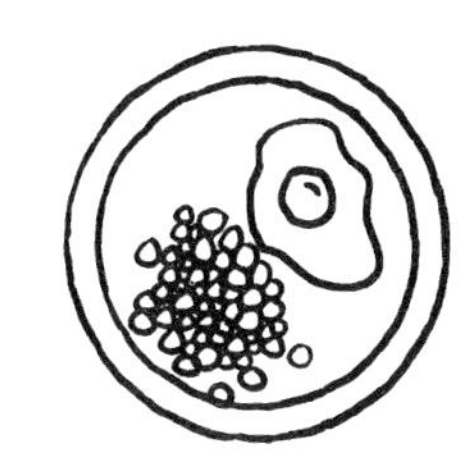

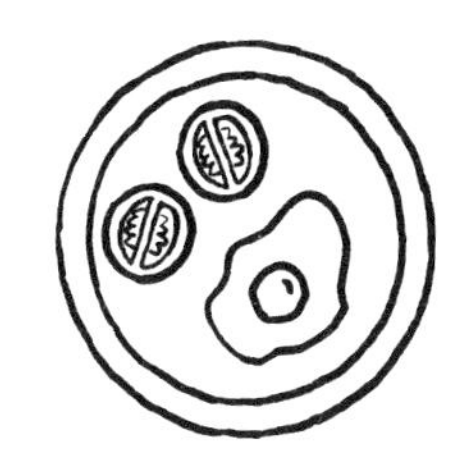

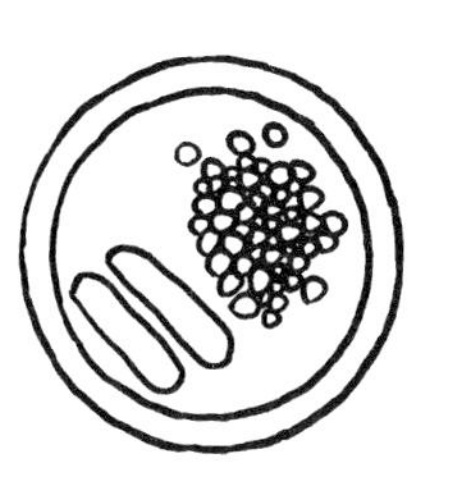

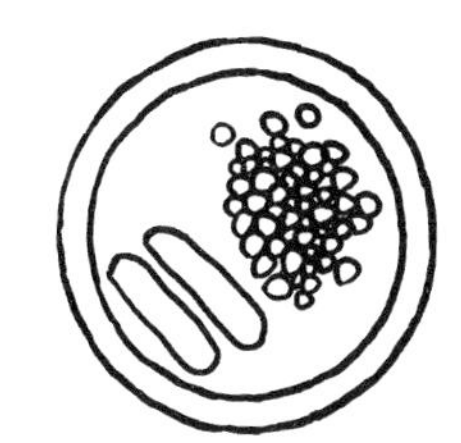
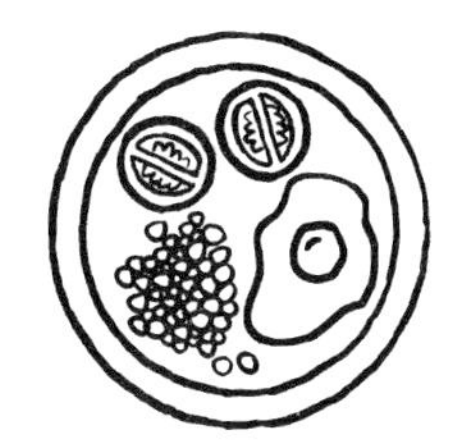

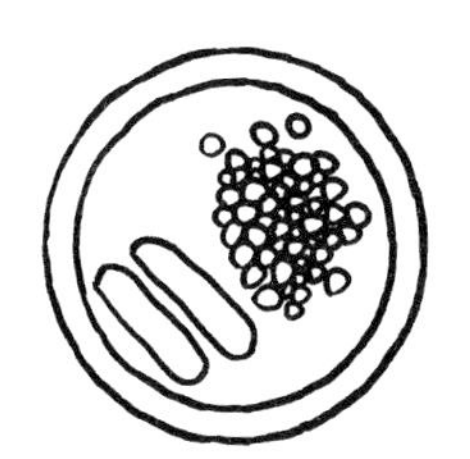

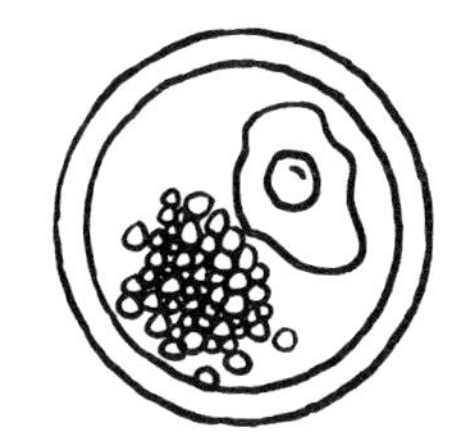

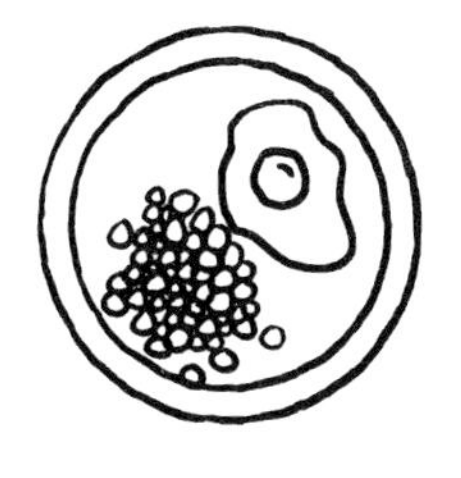

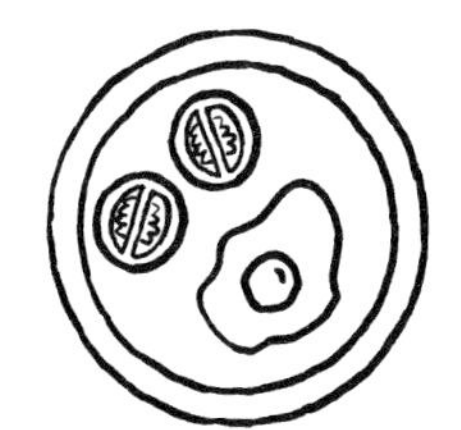

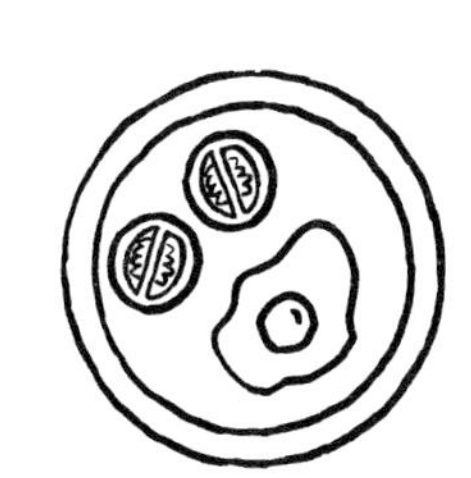

Worksheets

  

# Video shop



A new video shop will rent out videos, games and DVDs.

They do a survey to find out what people will borrow.

Each person fills out a form:

DVDs ✓ Games ✓ Videos ✓

DVDs ✗ Games ✗ Videos ✓

DVDs ✗ Games ✓ Videos ✗

DVDs ✓ Games ✓ Videos ✓

DVDs ✓ Games ✗ Videos ✓

DVDs ✓ Games ✗ Videos ✓

DVDs ✓ Games ✓ Videos ✓

DVDs ✓ Games ✓ Videos ✓

DVDs ✗ Games ✓ Videos ✓

DVDs ✓ Games ✗ Videos ✓

DVDs ✓ Games ✗ Videos ✓

DVDs ✓ Games ✗ Videos ✓

DVDs ✗ Games ✓ Videos ✓

DVDs ✓ Games ✓ Videos ✓

How many people will borrow **both** DVDs and games? ________________

How many people will borrow **both** DVDs and videos? ________________

How many people will borrow **both** games and videos? ________________



 

# Cars n colours

HD1/E2.1, HD1/E2.3

**Colour**

**Cars**

| | Red | Blue | Silver | White | Black |
|---|---|---|---|---|---|
| **Citroen** | 3 | 2 | 0 | 1 | 5 |
| **Fiat** | 1 | 3 | 1 | 2 | 4 |
| **Mini** | 1 | 4 | 0 | 0 | 2 |
| **Ford** | 0 | 3 | 1 | 2 | 2 |
| **Hyundai** | 2 | 1 | 1 | 0 | 3 |
| **Porsche** | 0 | 0 | 1 | 0 | 0 |

The table shows the results of a car survey.
Answer the following questions.

How many blue Fiats are there? ______________________

Are there any silver Minis? ______________________

What colour Ford is the most popular? ______________________

What colour Porsche is seen? ______________________

Which is the most popular coloured Hyundai? ______________________

How many white Fords are there? ______________________

How many black Hyundais are there? ______________________

Is it true that there are 6 blue Minis? ______________________

Are there any blue Citroens? ______________________

Worksheets

  

## Martial arts

HD1/E2.1, HD1/E2.3

Classes

Belts

| | Karate | Tae kwon-do | Jiu-jitsu | Judo | Kick boxing |
|---|---|---|---|---|---|
| White | 15 | 17 | 19 | 11 | 16 |
| Orange | 8 | 5 | 7 | 6 | 8 |
| Red | 6 | 4 | 1 | 5 | 3 |
| Blue | 2 | 6 | 2 | 3 | 3 |
| Green | 4 | 2 | 1 | 4 | 0 |
| Black | 2 | 1 | 2 | 2 | 1 |

Worksheets

How many white belts are there in the judo class? ____________________

In the kick boxing class how many black belts are there? ______________

How many orange belts are there in the jiu-jitsu class? _______________

How many green belts are there in the karate class? __________________

Which class doesn't have any green belts? __________________________

How many white belts are in the tae kwon-do class? ___________________

How many red belts are there in total? _____________________________

Which classes have one black belt in them? _________________________

 

## Record information



You can collect numerical information by **observing** things.

This is when you watch and record what you see.

Suppose you want to record the colours of the cars that go along the road in five minutes. The easiest way is using a **tally chart**.

Each time a car passes put a line next to its colour.

| Colour of car | Tally | Total |
|---|---|---|
| Red | 卌 卌 卌 \|\| | 17 |
| Black | | |
| Blue | | |
| White | | |
| Other colours | 卌 卌 | 10 |

Draw every 5th tally across the last 4 – this makes it easy to count the total.

In five minutes the following cars went past.

17 red cars  8 black cars  14 blue cars

12 white cars  10 other colours

The tally chart shows the red cars and the other colours.
Fill in the tallies and totals for the rest.

## Travel



Ask some students how they travel to college.

Record the results in the tally chart.

| Travel | Tally | Total |
|---|---|---|
| Cycle | | |
| Walk | | |
| Car | | |
| Bus | | |
| Train | | |
| Taxi | | |
| Other | | |

What is the most popular way of getting to college? ________________

Which way is the least popular? ________________

Did more students cycle or walk? ________________

Did more people catch the bus or the train? ________________

How many students took part in the survey? ________________





 

## Takeaway



Ask some people what their favourite takeaway is.
Record your results in the tally chart.
Fill in the total column.

| Takeaway | Tally | Total |
|---|---|---|
| Indian | | |
| Chinese | | |
| Fish and chips | | |
| Kebab | | |
| Pizza | | |

Do more people like Indian or Chinese best? ______________________

How many people like Chinese best? ______________________

How many people like pizza or kebab best? ______________________

Which is the most popular takeaway? ______________________

How many people did you ask altogether? ______________________

Should there have been an 'other' option? ______________________

Where did you ask people? ______________________

Do you think the results are reliable? ______________________

Why? ______________________



  

## Favourite shop



Here is a tally chart for you to fill in.

A survey asked people which shop they like best.
Their answers are listed below:

~~Locost~~ ~~Savo~~ ~~Locost~~ Savo Onestop Quicshop
Onestop Quicshop Other Locost Onestop Savo
Savo Quicshop Savo Other Savo Onestop
Locost Onestop Other Locost Quicshop Onestop
Savo Quicshop Other Quicshop Savo Onestop
Quicshop Onestop Savo Other Locost Other
Savo Quicshop Onestop Quicshop Other Onestop
Savo Locost Quicshop Onestop Savo Other
Quicshop Savo Onestop Savo

The first 3 answers have been put in the tally chart below.

**It is a good idea to cross out the answers as you use them.**

**Complete the tally chart.**

| Supermarket | Tally | Total |
|---|---|---|
| Locost | \|\| | |
| Quicshop | | |
| Onestop | | |
| Savo | \| | |
| Other | | |

## Soaps



A survey asked people to name their favourite soap.
Here are their answers:

| | | | |
|---|---|---|---|
| Brookside | EastEnders | Brookside | EastEnders |
| Other | Coronation St | Emmerdale | Coronation St |
| Emmerdale | Other | Brookside | Coronation St |
| EastEnders | EastEnders | Emmerdale | Emmerdale |
| EastEnders | Other | EastEnders | Coronation St |
| Brookside | Coronation St | Other | Brookside |
| Emmerdale | Other | EastEnders | Coronation St |
| EastEnders | Emmerdale | Coronation St | Emmerdale |
| EastEnders | Coronation St | Emmerdale | EastEnders |
| Coronation St | EastEnders | Other | Brookside |
| Other | EastEnders | Emmerdale | Coronation St |
| Emmerdale | Other | Coronation St | EastEnders |
| Brookside | EastEnders | Emmerdale | Coronation St |

**Complete the tally chart below.**

**Remember to cross out the answers as you use them.**

| Soap | Tally | Total |
|---|---|---|
| Brookside | | |
| Emmerdale | | |
| Coronation St | | |
| EastEnders | | |
| Other | | |



  

## Questionnaire



Here is a questionnaire about renting from the video shop.

Tick the boxes.

How old are you?

- Under 16 ☐
- 16-25 ☐
- 25-40 ☐
- 40+ ☐

How often do you rent items from a video shop?

- Once a week ☐
- More often ☐

Do you rent DVDs? ☐

Do you rent games? ☐

Are comedies your favourite rental? ☐

What type of game do you like best? ☐

**How could you improve this questionnaire?** ____________________

______________________________________________

______________________________________________

______________________________________________

______________________________________________

______________________________________________



 

## Travel to work



Suppose you want to know how people get to work.

**Circle the question you think is best.**

Do you come by car? Yes ☐ No ☐

Do you walk? Yes ☐ No ☐

How do you get to work? ______________________________

Have you caught a bus? Yes ☐ No ☐

**Circle the way you will ask questions.**

Verbally

Questionnaire

**Circle the type of people you will ask.**

Students

Adults

Children

**Circle the best time to ask questions.**

During the day | Late at night | Just after work

**Circle the best place to ask questions.**

In the car park | At the bus stop

On the train | At work



  

## Shopping lists

You can show more than one item by writing a times sign (×) and then the number you want.

**Example** 'Tin of paint × 3' means 3 tins of paint.

**Write a shopping list for these items.**

**1 cake**

**2 loaves of bread**

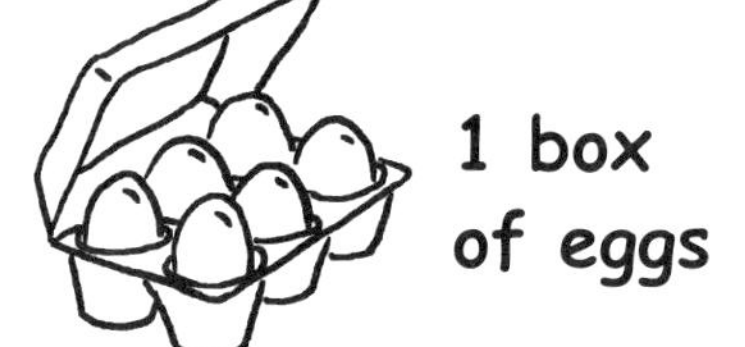

**1 box of eggs**

**4 cans of dog food**



**2 cartons of orange juice**

**1 bottle of coke**

**5 cans of beans**

**Shopping list**

___

___

___

___

___

___

___



 

# Decorating



Below is a shopping list.

Draw a picture in each box of one type of item.

Show how many of each type are needed.

A can of paint

Paint brushes × 4

Light bulbs × 3

Lamp shades × 2

Screwdriver



 

## Phone number list

HD1/E2.5

Order the list of names and numbers below so that your friend can find the phone numbers easily.

Think about how lists are ordered and why.

| | | | |
|---|---|---|---|
| Amy | 509012 | ________ | ____________________ |
| Wesley | 575890 | ________ | ____________________ |
| Rani | 622134 | ________ | ____________________ |
| Linda | 077715444356 | ________ | ____________________ |
| Avril | 770772 | ________ | ____________________ |
| Brian | 077755778827 | ________ | ____________________ |
| Jane | 574232 | ________ | ____________________ |
| Gary | 077790057867 | ________ | ____________________ |
| Sarah | 585423 | ________ | ____________________ |
| Vet | 501501 | ________ | ____________________ |

Worksheets

 

## Order the pages



You want your friend to look at the following pages in a book.

The pages will make more sense if they are read in the right order.

List the pages in each column in order.

| | | | |
|---|---|---|---|
| p. 15 | ________ | p. 46 | ________ |
| p. 24 | ________ | p. 91 | ________ |
| p. 11 | ________ | p. 27 | ________ |
| p. 9 | ________ | p. 31 | ________ |
| p. 2 | ________ | p. 68 | ________ |
| p. 8 | ________ | p. 34 | ________ |
| p. 17 | ________ | p. 75 | ________ |
| p. 1 | ________ | p. 56 | ________ |



  

## Cleaning rota



Jane, Roy and Mandy share a flat.

They have drawn up a rota of chores below.

Which day have they decided to have off? ________________

How many times do they vacuum per week? ________________

How many times do they polish and dust per week? ________________

| **Monday** | **Tuesday** | **Wednesday** | **Thursday** | **Friday** | **Saturday** |
|---|---|---|---|---|---|
| Wash up | Wash up | Wash up | Wash up | Wash up | Wash up |
| Polish and dust | Vacuum | Polish and dust | Vacuum | Polish and dust | Vacuum |
| Cook | Cook | Cook | Cook | Cook | Cook |

Choose a colour for each person:

**Key**

Jane ☐

Roy ☐

Mandy ☐

Colour code the rota so that the chores are split evenly between them.





 

## Cricket



The scores of each player in a cricket team's first two matches are listed below. Put the scores and each player's total in the table.
Don't forget a title and headings.

| | | |
|---|---|---|
| Joe Phillips | 20 runs | 57 runs |
| Frank Abbott | 15 runs | 12 runs |
| Connor Maude | 32 runs | 23 runs |
| Jalill Gakure | 37 runs | 16 runs |
| Jack Clarke | 22 runs | 15 runs |
| Brett Appleyard | 45 runs | 19 runs |
| Johan Mason | 9 runs | 21 runs |
| Ismael Annoot | 39 runs | 18 runs |
| Robert Moys | 12 runs | 18 runs |
| Rana Bakkan | 19 runs | 22 runs |
| Mark Stephens | 42 runs | 36 runs |

| | | | |
|---|---|---|---|
| | | | |
| | | | |
| | | | |
| | | | |
| | | | |
| | | | |
| | | | |
| | | | |
| | | | |
| | | | |
| | | | |



  

## Timetable

The times of the classes for your college courses are listed below.

| | | | |
|---|---|---|---|
| Monday | 10:00–12:00 Maths | 12:00–1:00 Lunch | |
| Tuesday | off all day | | |
| Wednesday | 9:00–11:00 Maths | 11:00–1:00 Lunch | 1:00–3:00 English |
| Thursday | 11:00–1:00 IT | 1:00–2:00 Lunch | 2:00–3:00 Maths |
| Friday | 11:00–12:00 IT | 12:00–1:00 Lunch | 1:00–3:00 English |

**Put this information into the timetable below.**

| | Monday | Tuesday | Wednesday | Thursday | Friday |
|---|---|---|---|---|---|
| 9–10 | | | | | |
| 10–11 | | | | | |
| 11–12 | | | | | |
| 12–1 | | | | | |
| 1–2 | | | | | |
| 2–3 | | | | | |



 

## Working in a shop



Information can be given in lots of ways.

Shops often colour-coded hangers to show sizes.

Below are some clothes with the sizes marked.

**Colour the key** – use a different colour for each size.

**Colour the hangers** to show the sizes.

**Key**

- ☐ Size 10
- ☐ Size 12
- ☐ Size 14
- ☐ Size 16

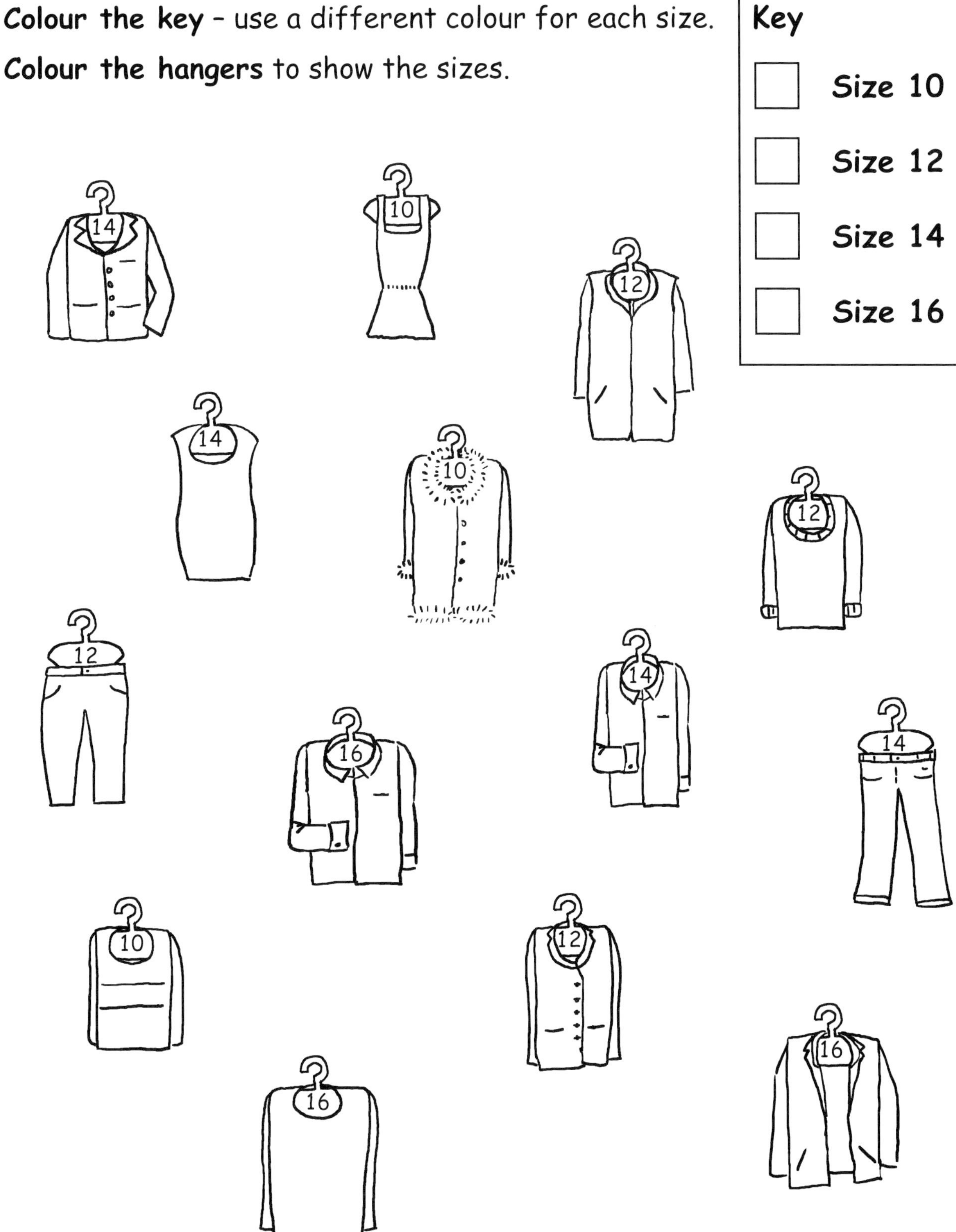

## Market day

Below is a plan of the market.

Use the following information to label the stalls.

- On one side there are three stalls.
  The middle one is the hot dog stand – label it **HD**.
- Opposite the vegetable stall is the butcher's stall – label it **B**.
- Next to the hot dog stand, on its left, is the toy stall – label it **T**.
- Opposite the toy stall is the fruit stall – label it **F**.
- Between the vegetable stall and the toy stall is the material stall – label it **M**.
- The largest stall sells shoes – label it **S**.
- The cake stall is next to the butcher's – label it **C**.
- Between the cake stall and the hot dog stall is the hardware stall – label it **HW**.

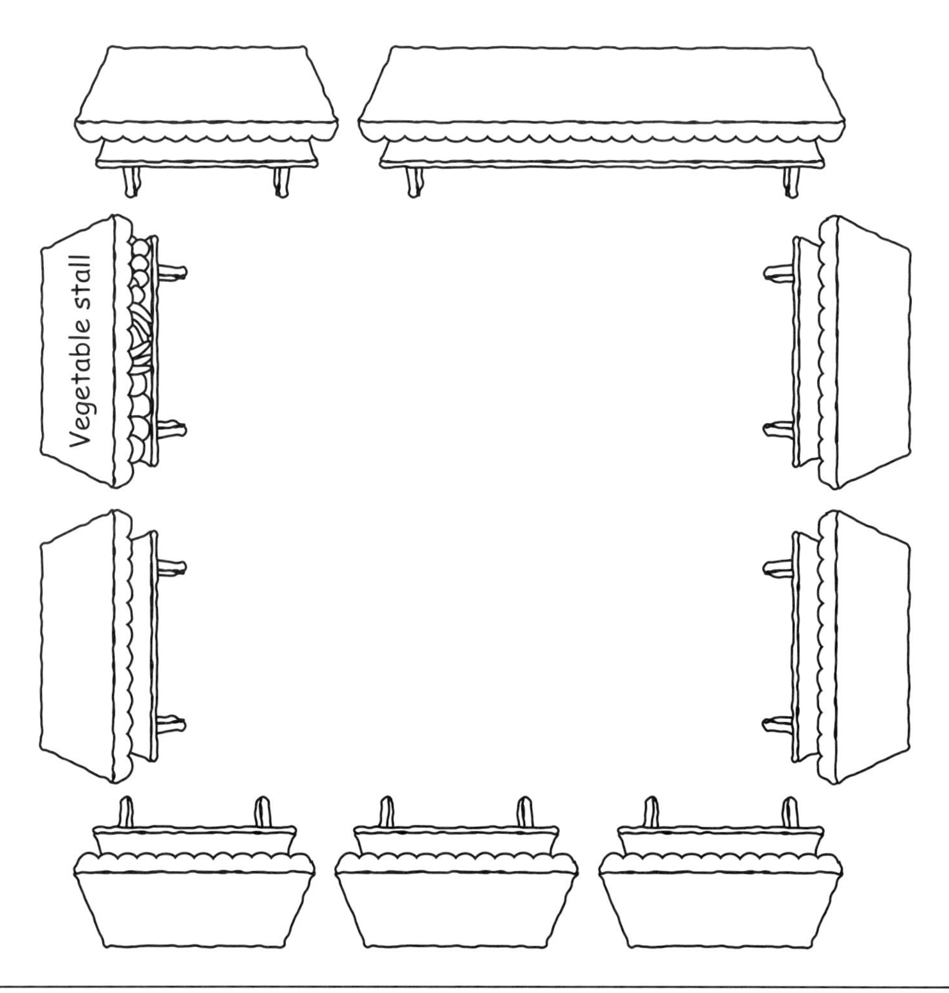





 

## Design a garden



Below is a plan of a house and garden.

Draw these on the grid:

- a **patio** in the shape of a **rectangle**
- **2 square flowerbeds**
- a **pond** in the shape of a **circle**.

Either label the items or use a key to show what they are.

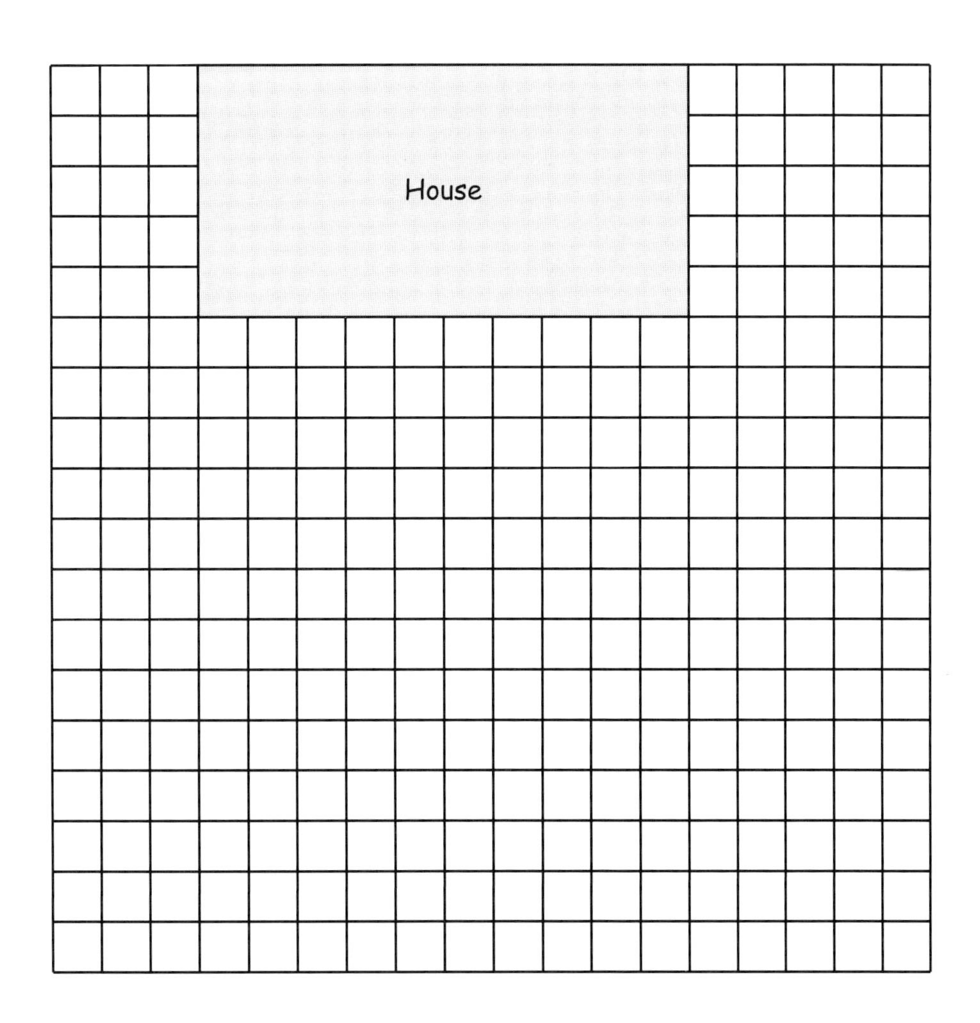

The college library colour-codes the books by subject.

**Maths** books have a **green** sticker.

**English** books have a **yellow** sticker.

**Geography** books have an **orange** sticker.

**Biology** books have a **red** sticker.

Mark the books below in the correct colour.

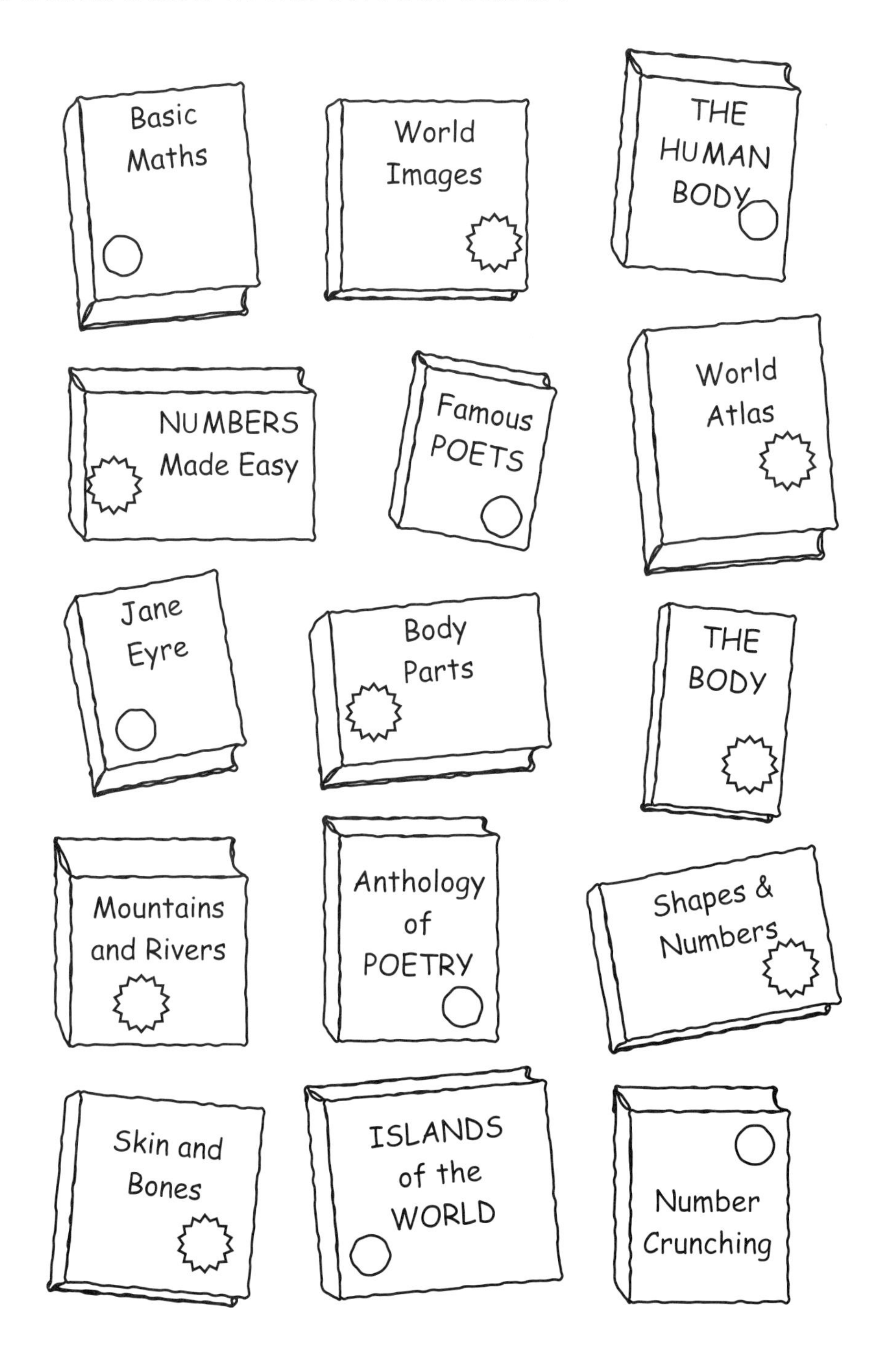




 

## Classroom plan

HD1/E2.1, HD1/E2.5

Here is a plan of a classroom.

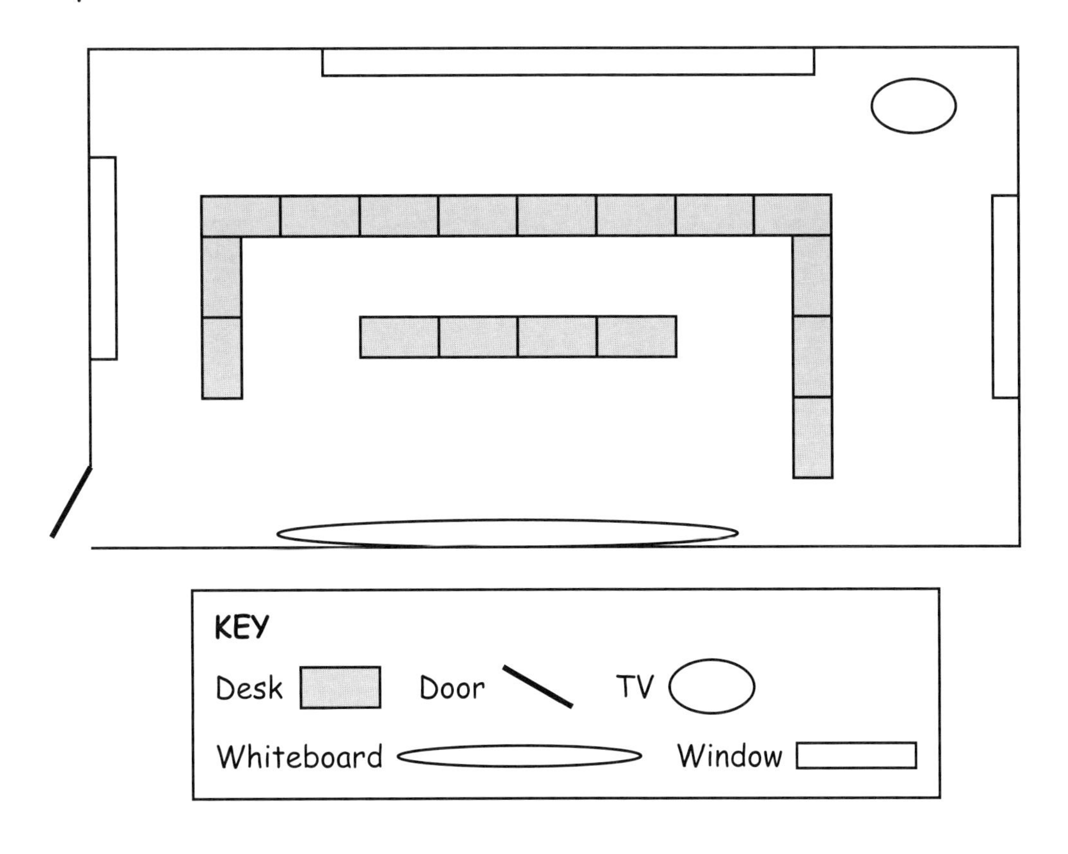


The key shows what the symbols mean.

How many doors are there? ______________

How many windows are there? ______________

How many desks are there? ______________

Draw a plan of your classroom.

First you need to sketch the shape of the room.

Look at the windows, doors and other items then draw them in.

Count the desks then draw them in.



  

## Room plans



**1** Draw a plan of your living room.

Show the chairs, sofa, TV and any other furniture you have.

Also show the doors and windows.

Write down 3 questions about your living room to ask somebody who has never been to your house.

______________________________________________

______________________________________________

______________________________________________

Could they answer these questions using your diagram? ___________

**2** Draw a plan of your kitchen.

Show the cooker, fridge, kitchen sink and any other items or units.

Also show the doors and windows.

Write down 3 questions about your kitchen to ask somebody who has never been to your house.

______________________________________________

______________________________________________

______________________________________________

Could they answer these questions using your diagram? ___________





 

# Vending machine

HD1/E1.3

Below is a tally chart. The number of tally marks give the total.
Fill in the rest of the totals

| Drinks | Tally | Total |
|---|---|---|
| Lemonade | ||| | 3 |
| Orangeade | 𝍸 | | 6 |
| Orange juice | 𝍸 | |
| Water | 𝍸 || | |
| Cola | 𝍸 |||| | |

Now draw a pictogram to show the results.

**Remember the title and key.**

Use  for each drink.

| | |
|---|---|
| **Lemonade** | |
| **Orangeade** | |
| **Orange juice** | |
| **Water** | |
| **Cola** | |

Worksheets

  

## Favourite cuppa

HD1/E1.3

Below is a tally chart. Fill in the totals.

| Favourite cuppa | Tally | Total |
|---|---|---|
| Tea | 𝍸 𝍷𝍷𝍷 | |
| Coffee | 𝍸 𝍸 | |
| Hot chocolate | 𝍷𝍷𝍷𝍷 | |
| Herbal tea | 𝍷𝍷 | |

Now draw a pictogram below to show the results.

**Remember the title and key.**

| | |
|---|---|
| | |
| | |
| | |
| | |

Worksheets

 

## Keep fit

HD1/E1.3

The tally chart shows what people do to keep fit.
Fill in the totals.

| Activity | Tally | Total |
|---|---|---|
| Yoga | \|\| | |
| Walk | ~~\|\|\|\|~~ \|\|\| | |
| Cycle | ~~\|\|\|\|~~ \| | |
| Run | ~~\|\|\|\|~~ | |
| Swim | ~~\|\|\|\|~~ \|\| | |

Draw a pictogram below.

**Remember the title and key.**

| | |
|---|---|
| | |
| | |
| | |
| | |
| | |

Worksheets

  

## Weather

The tables give information about the weather.

Draw a pictogram to show each set of data. Use squared paper.

### Snow

The number of times it snowed in the first 4 months of a year

| Month | Number of times it snowed |
| --- | --- |
| January | 7 |
| February | 8 |
| March | 4 |
| April | 1 |

**Remember the title and key.**

### Sun

The number of hours of sunshine in a week in June

| Day | Number of hours of sunshine |
| --- | --- |
| Monday | 6 |
| Tuesday | 10 |
| Wednesday | 7 |
| Thursday | 2 |
| Friday | 0 |
| Saturday | 5 |
| Sunday | 9 |

**Remember the title and key.**



 

## Watching TV

HD1/E2.5

A survey asked people how much time they spent watching TV in a week. Here are the results:

14 people spent less than 3 hours

16 people spent between 3 and 6 hours

21 people spent between 6 and 10 hours

25 people spent between 10 and 15 hours

30 people spent between 15 and 20 hours

28 people spent 20 hours or more.

Put this information into the table below.

| Hours spent watching TV | Number of people |
|---|---|
| | |
| | |
| | |
| | |
| | |
| | |

Now show the same information in a pictogram.

**Use squared paper and remember to include a title.**
**A pictogram has symbols – give a key to show what they mean.**

Worksheets

 

## Jewellery



A jewellery stall sold the following one day.
Record the data in the tally chart.

| | | | | | |
|---|---|---|---|---|---|
| bracelet | earrings | cuff links | necklace | earrings | brooch |
| necklace | watch | earrings | necklace | watch | bracelet |
| earrings | bracelet | watch | necklace | watch | brooch |
| cuff links | bracelet | cuff links | cuff links | earrings | earrings |
| necklace | watch | bracelet | brooch | earrings | earrings |
| bracelet | watch | earrings | brooch | bracelet | necklace |

| Jewellery | Tally | Total |
|---|---|---|
| Bracelets | | |
| Earrings | | |
| Cuff links | | |
| Necklaces | | |
| Brooches | | |
| Watches | | |

Now show this data in a bar chart – ask your tutor for some squared paper or graph paper.

**A bar chart uses a scale to show the number of items.**
**Remember to use a title and labels.**

Worksheets

 

## Surveys



The tables below give some survey results.
Discuss with your tutor the best way to display the data.
Use squared paper or graph paper for a pictogram or bar chart.

### Favourite keep fit activity

| Activity | Men | Women |
|---|---|---|
| Walking | 10 | 30 |
| Snooker | 30 | 5 |
| Cycling | 55 | 35 |
| Swimming | 30 | 30 |
| Football | 35 | 5 |
| Keep fit | 15 | 45 |
| Yoga | 5 | 25 |

### Favourite sweets

| Type of sweet | Number of people |
|---|---|
| Strawberry crème | 12 |
| Orange crème | 8 |
| Coffee crème | 15 |
| Fudge | 11 |
| Toffee | 6 |
| Nut surprise | 3 |

### Preferred type of car

| | Saloon | Estate | Sports car | Hatchback |
|---|---|---|---|---|
| Men | 5 | 45 | 55 | 15 |
| Women | 15 | 20 | 45 | 40 |

Worksheets

  

## Collect, record and display

Now collect your own information.

Here are some ideas that you could use in college.

You may find it easier to work in pairs.

- How many cars of each colour are there in the car park?
- How many people use the library at different times of day?
- How often do people go to the cinema?
- What is the most popular make of mobile phone?
- What is the most popular soft drink?
- How do people travel to college?
- What do people do at lunchtime?
- Are more people on part-time or full-time courses?
- What is the most popular type of TV programme?

Use your own ideas if you prefer.

For some ideas you need to observe things, for others you need to ask people questions.

Think carefully about how you will do this.

Write down any questions you will ask.
If you use a questionnaire, you may need to include 'other' boxes.
Ask a friend to try it out.

**Check your ideas with your tutor before you start.**



 

Draw at least two different charts/graphs to illustrate (show) your results.

At least one of your charts/graphs must be done by hand.

You may use a computer to do the other if you wish.

Colour in you charts/graphs and label them.

Write a few sentences about your survey.

Include answers to these questions - Why did you choose this topic?
- What did you find out?
- Did you enjoy it?

______________________________________________

______________________________________________

______________________________________________

______________________________________________

______________________________________________

______________________________________________

______________________________________________

______________________________________________

______________________________________________

______________________________________________

______________________________________________

______________________________________________

Worksheets